DATE DUE

632
Bla
00011256
Blanchard, Robert O
Field and Laboratory Guide to Tree
Pathology

Field and Laboratory Guide to Tree Pathology

Field and Laboratory Guide to TREE PATHOLOGY

ROBERT O. BLANCHARD

Department of Botany and Plant Pathology
University of New Hampshire
Durham, New Hampshire

TERRY A. TATTAR

Department of Plant Pathology
Shade Tree Laboratories
University of Massachusetts
Amherst, Massachusetts

1981

ACADEMIC PRESS
A Subsidiary of Harcourt Brace Jovanovich, Publishers
New York London Toronto Sydney San Francisco

Cover illustration adapted from photo courtesy of W.A. Stegall, C.S. Hodges, E. Ross, and E.G. Kuhlman, USDA Forest Service (see p. 168).

Cover design by Alma Hochhauser

ACADEMIC PRESS, INC.
111 Fifth Avenue, New York, New York 10003

United Kingdom Edition published by
ACADEMIC PRESS, INC. (LONDON) LTD.
24/28 Oval Road, London NW1 7DX

PRINTED IN THE UNITED STATES OF AMERICA

81 82 83 84 9 8 7 6 5 4 3 2 1

To Ellen and Donna

Contents

III Diseases Caused by Noninfectious Agents

IV Field and Laboratory Exercises

Preface

This Guide presents a selection of tree diseases, along with related field and laboratory activities, to provide the student with the basic information and skills necessary for tree disease diagnosis. In addition to students, we feel that professionals in arboriculture, forestry, landscape architecture, nursery practice, and other plant science specialties will also find it useful. Our selection of diseases is intended to be representative of the major health problems of forest and shade trees, with the belief that representative diseases understood well are of more value than all diseases understood superficially. The major emphasis is on tree diseases of the United States, although many aspects of the Guide are relevant to the study of tree diseases regardless of geographic location. We have divided the Guide into four Parts: Part I, Tools, Techniques, and Terminology; Part II, Infectious Diseases; Part III, Noninfectious Diseases; and Part IV, Field and Laboratory Exercises.

Although our emphasis is on recognition and diagnosis of tree diseases, we have included general control procedures for each disease. We have intentionally avoided recommendations of specific chemical controls, since these often change with time and locality.

We have used terms with the understanding that those that are useful, descriptive, and understandable will survive; those that are not will perish. Our terms are neither new nor universally accepted.

We have attempted to make our drawings as accurate and descriptive as possible. However, some structures and organisms may be slightly out of proportion due to artistic limitations and our attempt to achieve the best representations of entire disease cycles.

Suggested references are given and intended primarily as starting points for literature study. A selection of recent and older classic references are provided for most diseases treated after discussion of each disease.

Robert O. Blanchard
Terry A. Tattar

Acknowledgments

We would like to express our appreciation to a number of people who directly or indirectly contributed to the preparation of this Guide.

We thank Ms. Janis Larson for her artwork and Mr. Charles Shackett for printing of the illustrations. We thank Ms. Ellen Blanchard for her help in typing all the drafts of the manuscript. For their review of various parts of the manuscript, we thank Dr. George N. Agrios, Dr. Dale R. Bergdahl, Dr. A. Linn Bogle, Dr. Frank G. Hawksworth, Dr. Francis W. Holmes, Dr. David R. Houston, Dr. William A. MacDonald, Dr. Mark S. Mount, Mr. William Ostrofsky, Dr. Bruce R. Roberts, Dr. Richard A. Rohde, Dr. Walter C. Shortle, and Dr. Philip M. Wargo. To all those whose names appear in figure captions of photographs generously supplied, we are grateful. We also wish to acknowledge the assistance of the USDA Photo Library and the U.S. Forest Service Photo Library. Finally, we express special thanks to our families, colleagues, graduate students, and friends who have offered help, encouragement, and understanding.

PART I

Tools, Techniques, and Terminology

The goal of any course in tree pathology is to acquaint students with the principles, practices, and theories associated with the science and to give them sufficient background to diagnose tree diseases. Many diseases can be diagnosed with a high level of confidence in the field. However, most diagnosticians will make their final diagnoses in the laboratory. The complete tree pathologist must be *proficient* with many techniques associated with the science and must be *efficient* with the tools required for implementation of these techniques. Further, there must be common ground for communication between and among students, instructors, scientists, and the public. In this section we will present some basic information about the tools, techniques, and terminology associated with the study of tree diseases.

The Microscope

INTRODUCTION

Most organisms that cause diseases are very small, and a microscope is often necessary to see them. Proper use of the microscope plays an integral part in the identification of microorganisms. Generally, two types of microscopes are used in the laboratory as aids in the diagnosis of disease: (1) the compound microscope and (2) the dissecting or stereoscopic microscope.

COMPOUND MICROSCOPE

This rather delicate and complicated instrument employs a system of lenses so arranged as to give sharp, clear, and highly magnified images of very small transparent objects. Many types can be found in today's market, but the components described here are fundamental to all, regardless of the sophistication and complexity of various models.

Parts (Fig. 1.1)

1. Base—Firm support for the rest of the microscope.
2. Arm—For carrying the microscope and for support of the optical system.

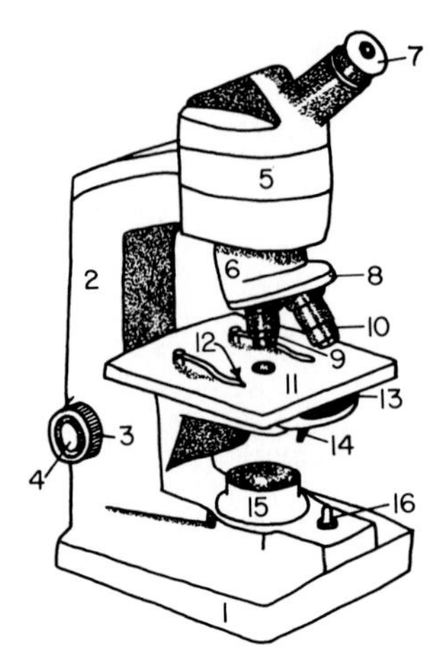

Fig. 1.1 Typical compound microscope.

3. Coarse adjustment knob—Moves the body tube and objectives to bring the specimen roughly into focus.
4. Fine adjustment knob—Brings the specimen into exact focus.
5. Upper body tube—Encloses light path between objective and ocular.
6. Lower body tube—Movable section for focusing objectives.
7. Eyepiece or ocular—Topmost optical lens system which further magnifies the primary image transmitted by the objective, frequently 10×.
8. Revolving nosepiece—Holds objectives which can be rotated into light path.
9. Low power objective—Lens system with initial magnification of 10×.
10. High power objective—Lens system with initial magnification of 43×. An oil immersion objective may also be attached to the nosepiece with initial magnification of 100×. This objective is used with oil between the objective and specimen.
11. Stage—Platform with an opening in the center on which is placed a slide containing the specimen to be observed.
12. Clamps—To hold the slide in place. May be a movable slide holder.
13. Condenser—Lens system which collects light rays and focuses them on the specimen.
14. Iris diaphragm—Assembly of thin metal leaves controllable by a lever to produce a variable sized opening.
15. Light source—Enclosed lamp with diffusion filter.
16. Light switch—For turning light source on and off.

Uses

The primary purpose of the compound microscope is to magnify a specimen that may be invisible to the naked eye. The amount of magnification depends on the lenses used and is roughly the product of the magnification of the ocular and the magnification of the objective. Thus, a 10× ocular and a 43× objective would yield a magnification of approximately 430 diameters. As a matter of practice, the following procedure should be used in bringing specimens up to the greatest magnification possible with the microscope:

1. Place a specimen slide on the stage, using the clamps to secure it in place.
2. Focus the low power objective on the specimen with the coarse adjustment knob. Then use the fine adjustment knob for exact focus.
3. Now place the high power objective into position, and again focus the specimen with the fine adjustment knob. Newer microscopes are constructed so that a specimen in focus with the low power objective will be approximately in focus with all objectives. Such lens systems are said to be *parfocal.*
4. Place a small droplet of immersion oil on the specimen slide, and bring the oil immersion objective into position. The oil serves to prevent bending of light rays out of the field and has the proper refractive index, dispersion, and viscosity characteristics for use between 100× objectives and specimen cover glasses. If the instrument is parfocal, the oil immersion objective may be turned into position without further adjustment. However, it is always good procedure to watch carefully to make sure the objective will not touch the slide. Focusing must always be done with the fine adjustment knob when using the oil immersion objective.
5. Regardless of the objective used, adjust the iris diaphragm to control contrast and definition of the specimen.

Alignment

For maximum effectiveness, the mechanical and optical components of the microscope must be clean and aligned. Follow this list of procedures each time a microscope is used. Some procedures may vary slightly with different microscopes, in which case the operation manual provided by the manufacturer should be consulted.

1. With the coarse adjustment knob, crank the body tube all the way up.
2. Clean each objective lens with lens paper (other papers will damage the delicate lenses) and a solvent to remove any oil or grease.
3. Clean the ocular lens.
4. Clean the slide stage.
5. Remove the light source.
6. Lower the condenser lens all the way down and clean it.
7. Make sure the lamp is not plugged in, and then open the lamp box and remove the diffusion filter. Clean the filter. Do not put fingers on the filter.
8. Close the iris diaphragm. To observe what this does, tip the microscope on its side and look through the condenser lens from the bottom (hold the

ocular lens so that it does not fall out). Open and close the iris diaphragm and observe what happens. Leave diaphragm closed.
9. While looking through the condenser lens, move the lens up and down. Note that the iris diaphragm goes in and out of focus.
10. Set the microscope back on the desk and put the lamp assembly back in place without the diffusion filter. Turn on lamp.
11. Place a piece of filter paper tightly over a slide and observe the image of the lamp filament on the paper. Do not look at the filament through the ocular. See if the filament is centered. Move the condenser lens up and down and observe the filament image on the filter paper. Leave the lens at a point where a peripheral ring of light begins to appear. At this point the image should be in focus on the filter paper. Now the light source will be focused on any object mounted for microscopic examination. This point is also the point at which the iris diaphragm was in focus when observing through the bottom of the condenser lens.
12. Do not change the position of the condenser lens after it has been set.
13. Replace the diffusion filter.
14. Put a slide with a specimen on the slide stage and secure with the stage clips.
15. With the low power objective lens in place, focus the specimen with the coarse and fine focusing knobs.
16. Open the iris diaphragm fully. While looking through the ocular, slowly close the diaphragm until the light begins to fade. Light from the condenser now just fills the back lens of the objective. This can also be accomplished by removing the ocular lens and observing the diaphragm through the open tube. Set the diaphragm at the point at which light just fills the lens.

Common Errors in the Use of the Microscope

1. Poor light adjustment—Control with iris diaphragm.
2. Dirty lenses—Both ocular and objectives.
3. Wrong objective—Do not use oil on high dry lenses. Do not use oil immersion lens without oil.
4. Inverted slide—Be sure the specimen is up.
5. Improper use of the coarse and fine adjustment knobs—Set the fine adjustment knob so that it can be turned in both directions before coarse focusing.
6. Never focus with the oil immersion lens first.
7. Never store the microscope with high dry or oil immersion lenses in place. Always store with low power lens in place.
8. Always clean immersion oil from the lens before storing the microscope.

DISSECTING OR STEREOSCOPIC MICROSCOPE

This instrument provides low power three-dimensional images of solid objects and low power images of transparent or translucent objects. Most student models range in magnifications from 10× to 50×. As with the compound microscope, many different types are available.

Parts (Fig. 1.2)

1. Base—Firm support for the rest of the microscope.
2. Stage—Flat surface for holding specimen.
3. Stage plate—Glass plate over opening in center. Illuminator beneath plate transmits light through transparent or translucent specimen.
4. Light switch—One on each side for control of illuminators.
5. Pillar—Support between base and movable parts.
6. Illuminator—Light source for transmitted or reflected light.
7. Focusing knob—Moves the body up and down to focus on a specimen.
8. Magnification knob—Allows continuously variable or set increment levels of magnification.
9. Eyepieces or oculars—Lens system which further magnifies the primary image transmitted by the objective lens.
10. Body—Movable component enclosing lens systems.
11. Objective lens—Lens system which supplies primary magnification of specimen.

Uses

The dissecting microscope is used for low power, three-dimensional magnification of opaque, transparent, or translucent specimens that are difficult to see with the naked eye. Typical uses might include observations of fungal fruiting structures in culture plates or on living tissues, or observations of disease symp-

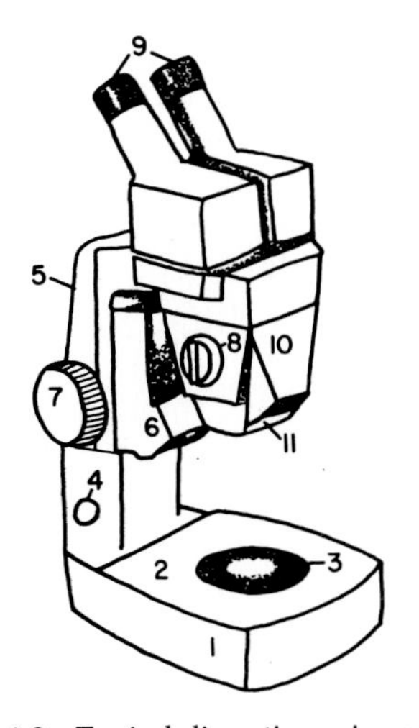

Fig. 1.2 Typical dissecting microscope.

toms on infected plant parts. Most dissecting microscopes allow low power magnification of transparent or translucent specimens such as leaves. The following procedures should be used when observing specimens:

1. Clean the stage plate, eyepieces, illuminator glasses, and objective lenses with lens paper.
2. Place a specimen or object in the center of the stage.
3. Illuminate the object (reflected light for opaque objects; transmitted light for transparent or translucent objects).
4. Raise the body above the focus point with the focusing knob. Then lower the body to bring the object into sharp focus.
5. Start with the magnification knob in the lowest position and adjust upward as necessary.

2

Sterile Technique

INTRODUCTION

Identification of microorganisms that cause tree diseases often requires microscopic examination. If a disease-causing microorganism has produced fruiting structures on the surface of diseased parts, observation of these structures and their accompanying spores with the microscope may provide adequate clues for identification. However, the presence of more than one type of fruiting structure may make it difficult to know which is the pathogen, or may make it impossible to obtain a pure sample of the pathogen for examination. Therefore, it is desirable to make pure cultures of pathogens for identification and subsequent storage for future reference.

Pure cultures are difficult to obtain and maintain, since propagules of bacteria and fungi can be found everywhere: in the air we breathe, the water we drink, the ground we walk on. Their ubiquitousness imposes the need for sterile technique. Sterile technique involves the use of procedures that will make and keep objects or materials free from living organisms other than a selected one. Hence, to obtain a pure culture of a pathogen, a propagule of the pathogen must be placed on a sterile substrate inside a sterile container. Sterile technique also requires that the transfer of the pathogen to the sterile substrate be done in such a way as to prevent contamination from other propagules of any kind.

PREPARATION OF STERILE MEDIA

All fungi and bacteria require specific elements and chemical compounds to grow and reproduce. Supposedly, if the right compounds in the right concentrations are provided, growth will occur. However, we have not found the right combinations to grow some microorganisms in the laboratory. On the other hand, many will grow very well in petri plates with some common natural media such as malt extract agar, potato dextrose agar, and corn meal agar. Some microorganisms may require supplemental materials such as vitamins before they will grow on these media.

When pure cultures of pathogens are desired, a suitable sterile medium must be prepared. Once the medium has been selected, the ingredients are measured into 1 liter flasks in an amount to yield 500 ml with distilled water (Fig. 2.1). The flasks should be no more than half full to prevent boiling over in the autoclave.

Flasks with medium are then plugged with cotton and placed in an autoclave at 15 pounds per square inch (psi) pressure and 121°C for 15–20 minutes. Autoclaving will dissolve and disperse the ingredients. This procedure is sufficient to sterilize the medium within the flasks. The medium is then allowed to cool slightly and is poured into disposable plastic petri dishes to a depth of about 5 mm. Certain techniques such as flaming the mouth of the flask before pouring

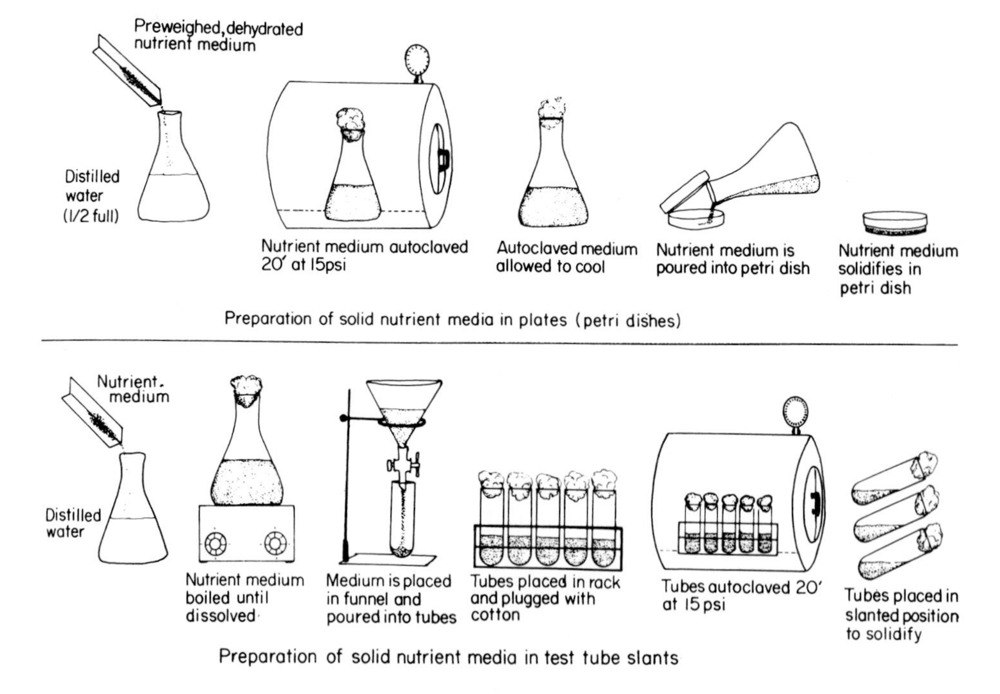

Fig. 2.1 Preparation of solid nutrient media in petri dishes and in test tube slants. (Drawing courtesy of George N. Agrios, "Plant Pathology," 2nd ed. Academic Press, New York, 1978.)

and pouring in a transfer chamber will help avoid contamination of the medium in the petri plates.

Once poured, the medium is allowed to cool until it gels. It is often a good idea to let the plates sit in a draft free area of the laboratory for a day or so to ensure that none have been contaminated. They then should be refrigerated until used.

Preparation of solid media in test tube slants is similar to that for petri plates (Fig. 2.1). The major difference is that the medium is heated to dissolve the ingredients and then poured into test tubes (about one third full). The test tubes are then plugged with cotton and autoclaved. When sterilization is completed, the tubes are placed in a slanted position until the medium solidifies. The

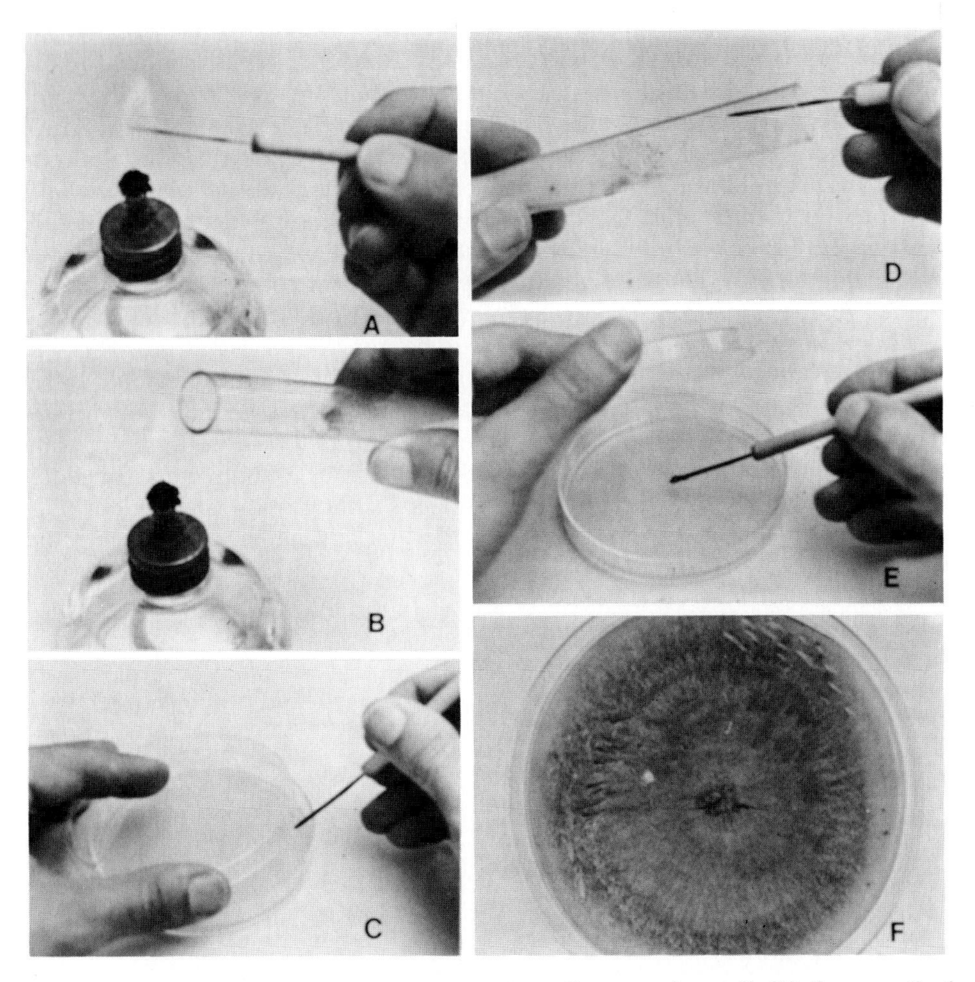

Fig. 2.2 Procedure for sterile transfer of a stock culture. Flame transfer needle (A), flame mouth of stock culture tube (B), cool transfer needle on sterile medium (C), remove sample of stock culture (D), place sample on sterile nutrient medium (E), incubate until growth is abundant (F).

slanted medium allows more surface area on which inoculated microorganisms can grow.

STERILE TRANSFERS

Fungi and bacteria will not grow indefinitely on media in petri plates or on agar slants in test tubes. Therefore, periodic transfers of the microorganisms should be made to new media to keep them viable. Many cultures have been lost because of improper transfer techniques.

Once a suitable medium has been prepared and sterilized (Fig. 2.1), transfers can begin. It is best to do transfers in a transfer chamber, but if one is not available, surface sterilize the work area with a 20%–20%–60% alchohol–sodium hypochlorite bleach–distilled water solution or other suitable disinfectant. Flame sterilize a transfer needle or loop to redness along its entire length (Fig. 2.2A). This is important because no known organism has demonstrated an ability to survive burning. However, spores attached to portions of the needle which have not come in contact with the flame, may survive the transient heat applied. Remove the cap or cotton plug from the stock culture tube and lightly flame the mouth (Fig. 2.2B). This will kill any propagules of microorganisms that are in contact with the glass. Cool the transfer needle by touching it briefly to the sterile medium on which the organism is to be transferred (Fig. 2.2C). This will ensure that residual heat in the flamed needle will not kill the sample being transferred. Remove a sample of the organism from the stock culture with the transfer needle (Fig. 2.2D) and place it in a new culture tube or on a culture plate (Fig. 2.2E). Reflame the mouth of the stock culture tube and the transfer needle. Incubate the newly transferred microorganisms until growth is abundant (Fig. 2.2F). New stock cultures can then be placed in a refrigerator for storage. Cold temperatures, (approximately 5°C), as in a refrigerator, will not kill the organisms, but will substantially reduce their metabolism. Microorganisms vary in the amount of time they can survive cold storage.

3

Isolation of Disease-Causing Organisms

GENERAL

Diagnosis of a particular disease requires identification of the pathogenic agent. If signs of the pathogen, such as fruiting structures, are present, identification may be made by preparing a slide directly as described in Chapter 4. However, if only symptoms are present, isolation of the pathogen is required. The diagnostician is often placed in a difficult position at this point, because more than one causal agent can produce identical symptoms. Furthermore, not all causative agents are biotic, and of those that are biotic, not all can be isolated in a culture medium. Viruses, mycoplasmalike organisms (MLO), and many fungi, particularly rusts and powdery mildews, will not grow in common culture media. Fortunately, most rusts and powdery mildews produce identifiable fruiting structures on the infected hosts.

Essentially, isolation of disease-causing organisms is restricted to bacteria and fungi, but a few important facts must be understood before proceeding to a description of actual isolation. Pathogens become pathogens because they have successfully penetrated host tissues. Therefore, a pathogen that is inside the infected part may not have developed any visible fruiting structures or cells. If symptoms are of the necrotic type (see Symptoms and Signs, Chapter 7), the pathogen may still be in the dead tissues, but so may many saprophytic organisms. Tissues that are killed by a primary causal agent are soon colonized by a variety of nonpathogenic organisms and a distinction must be made between the two. Therefore, dead tissues should be avoided. The pathogen, by itself, is most likely to be found at the margin of a lesion, in the transition tissues between healthy and dead tissues. Many surface contaminants may also be present on these transition tissues, so surface sterilization is essential. Use of fresh tissue is also critical since most pathogens are soon overwhelmed by saprophytes shortly after tissues are cut from the tree.

Pathogens that produce necrotic symptoms, such as wilt, or those that produce nonnecrotic symptoms, such as tumefaction or fasciculation, may be difficult to isolate. For example, wilted leaves may result from a root pathogen which is far removed from the leaves. But, generally, properly selected and surface sterilized tissues will yield nearly pure cultures of a culturable pathogen.

ISOLATION OF FUNGI

Fungal pathogens may be isolated from leaves as illustrated in Fig. 3.1. If the infected part is a leaf, 5–10 mm^2 sections are cut from the margin of the lesion and placed in a sterilizing solution. Several sections should be used so that the

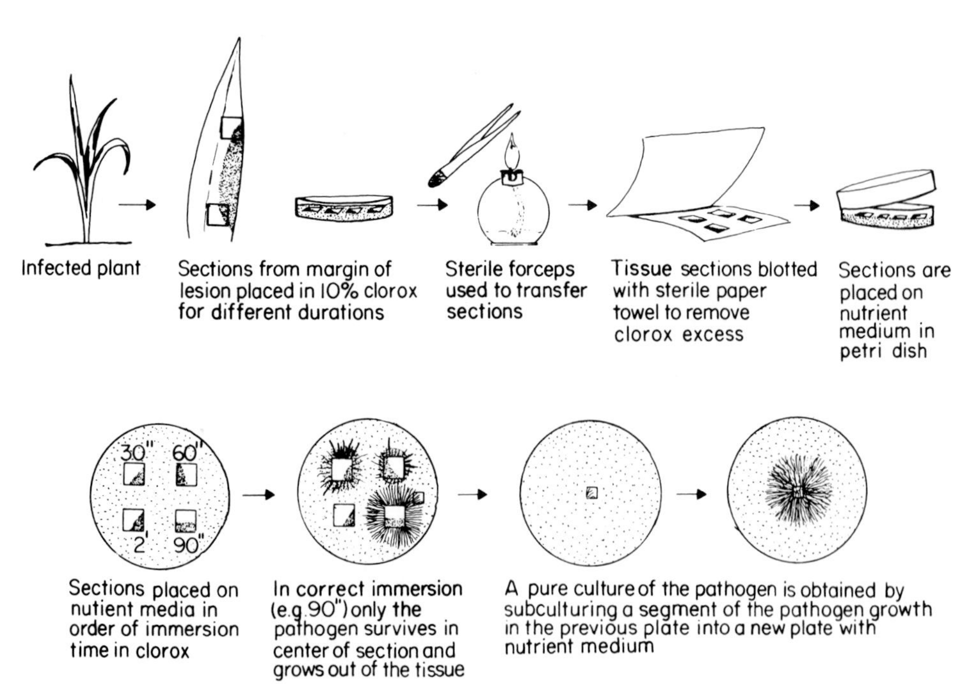

Fig. 3.1 Isolation of fungal pathogens from infected leaf tissue. (Drawing courtesy of George N. Agrios, "Plant Pathology," 2nd ed. Academic Press, New York, 1978.)

duration in the sterilizing solution can be varied. This will increase the chances of obtaining a section free of contaminants. Sections are then removed with sterile forceps and blotted dry on sterile paper. If they are put directly on nutrient medium without blotting, the excess sterilizing solution may diffuse into the medium and prevent germination or growth of pathogen spores or hyphae. Sections are then placed on nutrient medium and incubated. Those that yield only one fungus are likely to yield the pathogen. A small portion of the fungal growth from this section can be transferred to a new plate of culture medium, producing a pure culture.

If the infected part is a woody stem or root, the affected tissues are surface sterilized by wiping with a swab saturated in sterilizing solution. Small sections of bark and/or underlying wood at the margin of a section are aseptically removed and placed on nutrient medium (see Exercise IV for details). Procedures are then the same as for leaf sections.

ISOLATION OF BACTERIA

Pathogens that are suspected as being bacterial may be isolated from leaves as illustrated in Fig. 3.2. Procedures are similar to those for isolation of fungi except that instead of putting host sections directly on nutrient medium, the sections

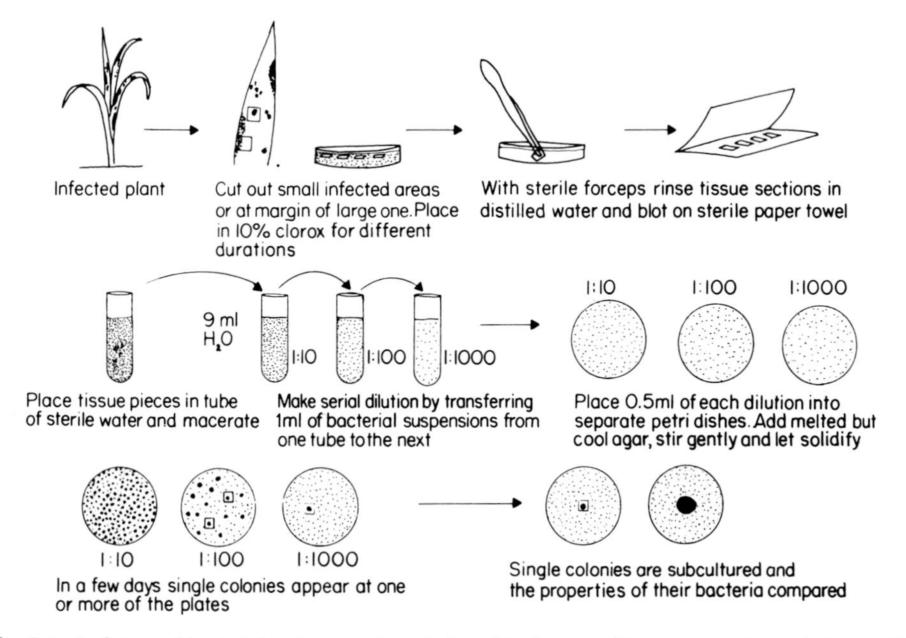

Fig. 3.2 Isolation of bacterial pathogens from infected leaf tissue. (Drawing courtesy of George N. Agrios, "Plant Pathology," 2nd ed. Academic Press, New York, 1978.)

are macerated in sterile water. Serial dilutions of the suspension are made, and 0.5 ml of each dilution are spread over nutrient medium.

On some occasions, if a particular pathogen is suspected but verification is required, sections of diseased tissues may be placed on a piece of wet filter paper in a sterile petri dish. This is commonly known as a moist chamber. Under such conditions the pathogen may produce identifying fruiting structures or spores on the surface of the sections.

Selected Reference Films

1. "Isolating Pathogenic Fungi."
2. "Isolating Pathogenic Bacteria."
Both films are available from APS Headquarters, 3340 Pilot Knob Road, St. Paul, Minnesota.

Preparation and Interpretation of Microscope Slides

INTRODUCTION

A tree disease can often be accurately diagnosed by macroscopic examination for characteristic symptoms of the disease. Ultimately, however, the microscope must be used for positive identification of a particular pathogen, and the best microscope available is of no use if a sample is improperly prepared for examination. The desired result of the examination will dictate the procedures used in preparing the material.

Simple identifications may require no more than placing a loopfull of spores in a drop of water and covering them with a cover glass. Other identifications may require examination of hyphae or structures on which spores are borne. Still others may require that the organisms be treated with specific stains to emphasize obscure structures. Then, once the organism has been identified, preservation on microscope slides may be desirable for future reference. Use of a few basic techniques will usually enable the examiner to prepare, identify, and preserve organisms that cause tree diseases.

MOUNTING MATERIAL FOR EXAMINATION

First, a sample of the organism must be obtained. If the pathogen produces fruiting structures on the leaf, stem, or other tree part, these structures can be placed on microscope slides for identification directly. However, if the pathogen is beneath the surface tissues and produces only mycelium, pure culture isolations must be done (these are described in Chapter 3).

Place a drop of water on a glass slide (Fig. 4.1A). The amount can be determined with practice, but should be enough to fill the area under the applied cover glass, and not so much as to cause the cover glass to move when jarred. Add a portion of the sample to be examined (Fig. 4.1B) and tease it apart with two inoculation needles (Fig. 4.1C). Remove excess large pieces. If the cover glass does not lie flat when applied, too much material has been added.

Place the edge of a cover glass on the slide at a slight angle and bring it in contact with the drop of water (Fig. 4.1D). Hold the cover glass up with an inoculation needle and then slowly withdraw the needle toward the end of the

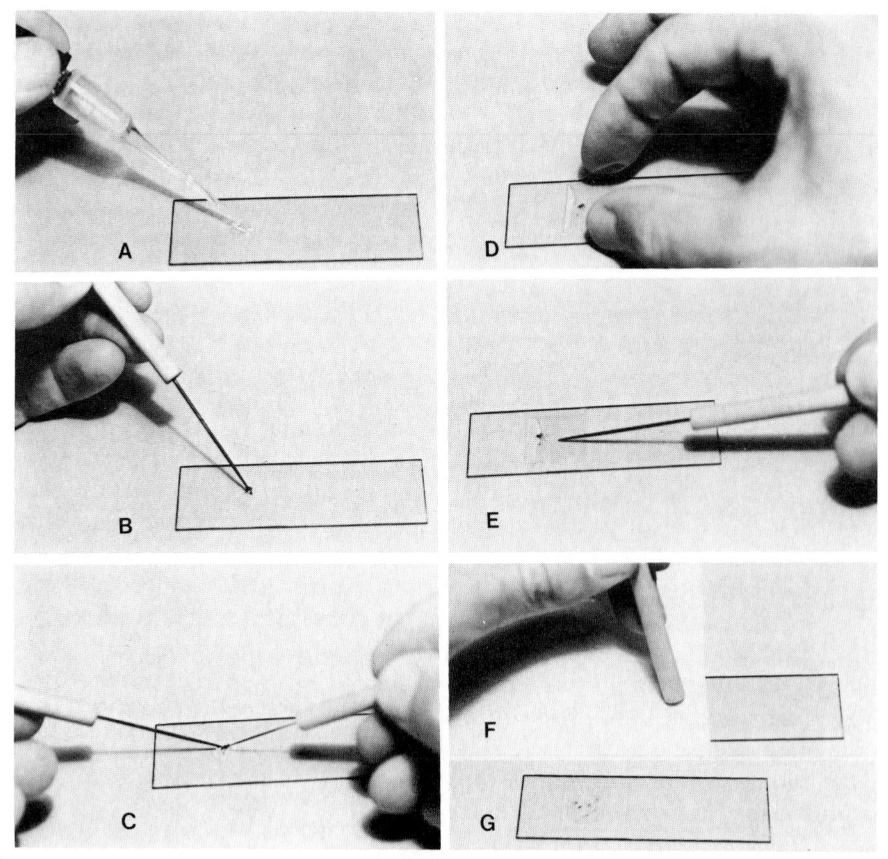

Fig. 4.1 Procedure for mounting material on glass slides for microscopic examination. Place a drop of water on slide (A), add sample (B), tease apart (C), touch edge of cover glass to edge of water drop (D), hold cover glass with needle and withdraw slowly (E), tap absorbent paper over cover glass to remove excess water (F), material ready for examination (G).

slide (Fig. 4.1E). This procedure will prevent excess air bubbles from being trapped under the cover glass.

Cover the slide and cover glass with a piece of absorbant paper and lightly tap the area over the cover glass with the eraser end of a pencil or the wooden handle of an inoculation needle (Fig. 4.1F). This will remove excess water and flatten the sample into a workable optical range. Now the slide is ready for examination (Fig. 4.1G). Lactophenol, with or without stain, can be used in place of water.

FIXATION AND STAINING

Alterations in the structure of an organism begin immediately after death. Fixation is a process that prevents or limits these alterations so that the appearance of the organism after fixation is similar to its appearance in the living state. In other words, the living structures are "fixed" in position. For identification purposes, it may not be necessary to fix an organism, but if staining and preservation are desired, it is a useful technique. Bacteria and fungi are usually placed live in fixatives such as acids, alcohols, and phenols.

The type of fixative used depends on the results desired. For examination of external morphological features, any fixative that will kill the organism without causing plasmolysis may be used. However, if examination of structures within cells is desired, or if thin sections of large fruiting structures must be prepared, more precise techniques are required. Such techniques may require dehydration (removal of water from cells), embedding (infiltration of tissues with a medium such as paraffin, which will allow the tissue to be cut sufficiently thin for the microscope), and staining (increasing contrast in transparent structures).

Stains are chemical compounds that selectively bind to specific sites within cells. In this way, they provide contrast to the bound sites and are useful in identifying the presence and location of cellular constituents. For example, some stains are selective for DNA and will highlight nuclei. Others may be selective for chitin and will give contrast to cell walls. Still others may be specific for enhancing flagella.

Fixatives and stains are usually used together and therefore must be compatible. Selection of an adequate fixative, combined with a suitable stain, will lead to the enhancement of cellular and tissue structures. Organisms prepared in this way are ready for microscopic examination and hopefully for identification and preservation.

SEMIPERMANENT AND PERMANENT MICROSCOPIC MOUNTS

Eventual reexamination of a microscopic specimen and verification of its identity is possible only if it has been adequately preserved on a microscope slide. If preservation is required for less than a year, then a semipermanent, nonhardening mount may be adequate. If indefinite preservation is required, then a permanent, hardened mount should be prepared.

Semipermanent mounts can be made from various formulations, but one of the best preparations for fungi is lactophenol. This compound is made by heating the following ingredients in a hot water bath:

Phenol	20 g
Lactic acid	20 g
Glycerin	40 ml
Distilled water	20 ml

Stains can be added to the lactophenol if desired. When a suitable specimen has been placed in lactophenol on a slide, a cover glass is added as in Fig. 4.1. The slides should be dried for several weeks at about 50°C and then sealed by ringing the edges of the cover glass with fingernail polish. Sealing is required, because although the preparation has been dried, the mounting medium will remain liquid and evaporate in time. The fingernail polish will also deteriorate with age.

Permanent mounts will last indefintely if properly prepared. The procedure is similar to that for semipermanent mounts except that mountant will harden, thus minimizing deterioration due to evaporation. Several prepared mounting media such as balsam are available from chemical supply houses; others can be prepared in the laboratory. The beginning student in tree pathology will most often find water mounts and semipermanent mounts adequate.

INTERPRETATION

Interpretation of material on a microscope slide often requires more than just looking at the slide. What we see in the microscope may not be an accurate representation of what exists in nature. Several factors may contribute to distortions in what is real, and the microscopist is forced to "interpret" that which is seen. For example, a fungus isolated on a nutrient medium, which provides an unlimited food source and a lack of competition, may produce structures and assume configurations that do not appear in nature. The fungus, or parts of the fungus may then be placed on a microscope slide. If placed in water so that it is still alive, chances of further distortion are minimized. However, if the fungus is killed in a fixative and stained, chances for distortion are increased. Further distortions are likely if the fungus is dehydrated, embedded, and sectioned. The microscopist must be aware of these potential sources of artifacts and distortions so that accurate interpretations can be made.

Whole mounts of fungi are probably the easiest to interpret. However, several slides should be made from several locations of a culture to ensure that as many structures as possible are viewed. Even then, understanding of how spores are borne may be obscured, since spores of many fungi become dislodged when immersed in mounting liquid. In this case the fungus should be examined directly on the culture plate with a dissecting microscope or with the low power lens of a compound microscope. Interpretations become easier when examinations are made at more than one level of magnification.

Sectioned fungal material becomes less easy to interpret. First of all, to be sectioned, the specimen has to pass through several steps of preparation, any one of which may introduce distortion. Second, sections of the specimen will be very thin, and it is often necessary to know something about the relationship of the section to the whole specimen. A section through the center of a golf ball can be taken at any angle without misrepresenting its structure. However, a section through the center of an elliptic football will represent structures that look different at each angle sectioned.

A third type of material requiring interpretation is sectioned host tissue with embedded or superficial pathogen structures. The untrained eye may spend an entire laboratory period looking at healthy host tissue and miss the host–pathogen association. Therefore, it is important when examining the affected portion of a tree in the field or in the laboratory to know which structures or anatomical features are part of the healthy tree and which features are either

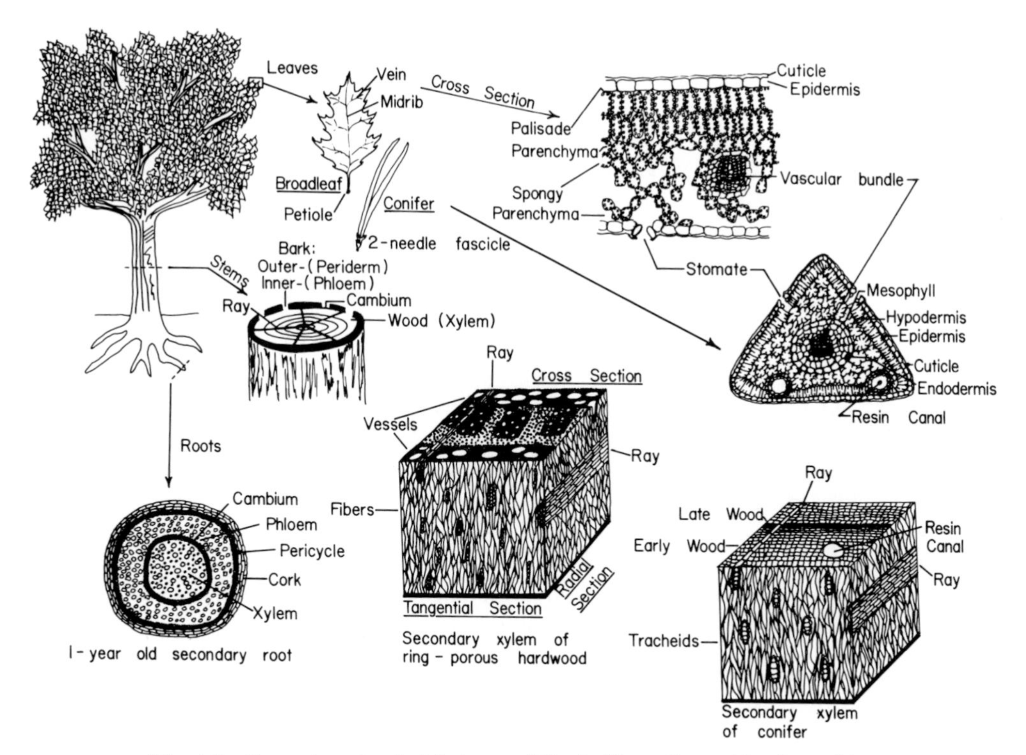

Fig. 4.2 General anatomical features of "typical" conifer and hardwood trees.

caused by or are part of the pathogen. Figure 4.2 gives a brief review of the general anatomical features of "typical" conifer and hardwood trees. It may be helpful to refer to this figure when viewing sections of diseased tissue through the microscope. Remember, sections are thin, and their orientation is important. Sectioned material will exhibit only two-dimensional images of three-dimensional structures. Know the sectioning angle and, if possible, observe several sections at various magnifications before attempting interpretation.

Successful interpretation is aided by accumulated knowledge and careful observation. No one disputes the fact that the identification of a pathogen is easier when one knows what it is. Interpretation is no different, and the job is easier when one takes the time to learn the limitations of tissue preparation and to understand the need for observation of several samples.

Selected References

Core, H. A., Cote, W. A., and Day, A. C. (1979). "Wood Structure and Identification," 3rd ed. Syracuse Univ. Press, Syracuse, New York.

Esau, K. (1967). "Plant Anatomy," 2nd ed. Wiley, New York.

5

Preservation of Diseased Specimens and Pathogens

INTRODUCTION

Preservation in tree pathology implies the killing of or stopping the growth of microorganisms associated with the disease. If the microorganism is within a host, preservation of the host tissues may also be desired. A major problem is the lack of a specific and convenient technique that can be used successfully for all microorganisms and host tissues. Furthermore, many pathogens lose their pathogenicity after several transfers. It is therefore necessary to use various techniques that are designed for preservation of specific microorganisms and plant parts. The desired preserved condition, whether living or killed, will dictate the method used. Below are descriptions of some of the most common methods of preservation used by tree pathologists.

DISEASED SAMPLES

1. Plant Press (Fig. 5.1)

A plant press is particularly useful when preserving infected leaves. Leaves that are still alive must be dried quickly to prevent saprophytic growth of molds. Another important consideration is to preserve the leaf, symptoms of the dis-

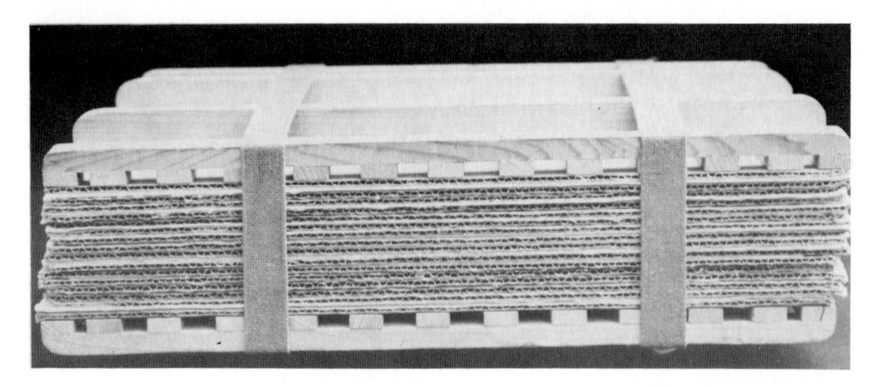

Fig. 5.1 Plant press for drying leaf samples.

ease, and any structures of the pathogen in a state as near natural as possible. Leaves should be placed flat between sheets of newspaper. The newspaper is inserted between two blotters which are inserted between two pieces of corrugated cardboard. This makes up one unit and several units may be strapped together between rigid wooden covers. The whole plant press is then placed on its side in a ventilated oven or over a heat source consisting of two 100 watt light bulbs. As soon as the material has dried, it may be mounted on herbarium paper or stored in suitable containers.

2. Riker Mounts (Fig. 5.2)

Samples of leaves dried in a plant press or samples of small dried twigs and branches exhibiting symptoms and signs can be preserved in Riker mounts. These can be purchased from most biological supply houses. They consist of a thin box filled with cotton and sealed with a glassed cover. The specimen is

Fig. 5.2 Riker mount for preservation of dried leaves, twigs, or branches.

arranged for easy viewing on the cotton. The glassed cover is then put on the box and secured with tape or needles. Specimens preserved in this way are useful for demonstration purposes.

3. Pickling (Fig. 5.3)

Diseased leaves, twigs, and fruits, as well as fruiting structures of some fungi, can be preserved in liquids that "pickle" the specimens. Weak acids, alcohols, and formaldehyde are a few of the pickling agents used. Specimens are easily viewed through the clear glass containers. One disadvantage to this technique is that the natural color of the specimen is often bleached out.

4. Air Drying

Woody specimens, such as infected twigs, branches, or roots can usually be preserved intact by air drying. Drying cracks may develop, depending on the tree species, but disease symptoms can still be recognized.

Fig. 5.3 Diseased sample preserved in pickling agent.

PATHOGENS—FUNGI AND BACTERIA

1. Agar Slants (Fig. 5.4)

Agar slants are used for preservation of pathogens in a viable but nongrowing state. Propagules are inoculated onto agar slants (see Chapter 2) and allowed to nearly cover the surface. They are then refrigerated, causing growth of the pathogen to slow down or cease. Periodic transfers must be made, depending on the species.

2. Mineral Oil

Pathogens are grown on agar slants as described in the previous paragraph. Slants are then covered to a depth of 1 cm above the uppermost edge with sterilized mineral oil. Tubes may be stored at room temperature but are best kept in a refrigerator. Cultures will remain viable for up to two years or more depending on the species. Two advantages of this technique over simple agar slant storage are that some species remain viable longer, and it is a simple method for controlling mites.

3. Freeze-Drying (Fig. 5.5)

Freeze-drying or lyophilization has become one of the most popular methods of preservation, because bacteria and fungi can be preserved in a viable state for long periods of time. The process involves freezing the spores or cells in a colloid such as skim milk or in a sugar solution. The sample is then dried from the frozen state and sealed under vacuum in ampuls. Ampuls are best stored under refrigeration.

Freeze-drying is also a convenient and practical method for preserving nonviable fruiting bodies, petri plate cultures, and diseased specimens. Specimens are frozen and then placed directly in a freeze-drying apparatus. When dry, the

Fig. 5.4 Agar slant with viable fungal pathogen on the surface.

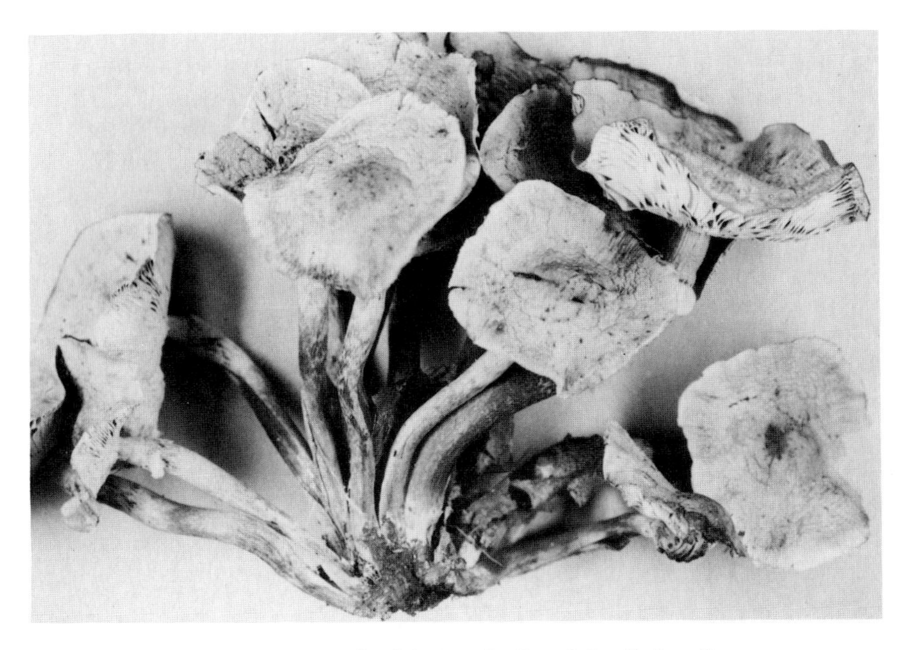

Fig. 5.5 Freeze-dried fruiting bodies of *Armillaria mellea.*

specimens can be placed in boxes or other containers for future examination. Specimens retain their natural shape and staining characteristics. Colors often remain nearly true.

4. Silica Gel (Fig. 5.6)

This technique has proven useful in maintaining stock cultures of fungi in viable but inert condition for several years. An advantage in this method, as well as in freeze drying and storing under vacuum, is that pathogens are less likely to

Fig. 5.6 Test tube containing viable spores of a fungal pathogen on silica gel particles.

lose their pathogenicity. Also, the chance of initiating cultural mutants is minimized. The procedure involves adding viable spores to 2-3 ml of sterile non-fat dry milk and then distributing about 1 ml of the suspension over frozen half-filled tubes of silica gel (6-12 mesh, grade 40). Screw-cap test tubes work best. Storage should be in a refrigerator. New cultures can be started by shaking a few silica gel particles from a test tube onto sterile medium. The tube with silica gel can then be returned to the refrigerator for future transfers. This procedure has a distinct advantage over preservation under vacuum in ampuls, because once an ampul is opened, the entire contents should be used immediately.

5. Air Drying

Air drying is primarily used for preservation of large fruiting structures of fungi such as conks and mushrooms. Conks do not lose their shape, but mushrooms generally collapse and become distorted. However, carefully dried mushrooms will still retain microscopic characteristics which are diagnostic for identification. Only fresh healthy specimens should be used because infested and infected specimens may deteriorate before they can be dried.

Selected References

American Type Culture Collection (1980). "Laboratory Manual on Preservation, Freezing, and Freeze-Drying, As Applied to Algae, Bacteria, Fungi and Protozoa." Rockville, Maryland.

Barratt, R. W., and Tatum, E. L. (1950). A simplified method for lyophilizing microorganisms. *Science* **112,** 122–123.

Buell, C. B., and Weston, W. H. (1947). Application of the mineral oil conservation method to maintaining collections of fungous cultures. *Am. J. Bot.* **34,** 555–561.

Ellis, J. J. (1979). Preserving fungus strains in sterile water. *Mycologia* **71,** 1072–1075.

Hanlin, R. T. (1972). Preservation of fungi by freeze-drying. *Bull. Torrey Bot. Club* **99,** 23–27.

Kohlmeyer, J., and Kohlmeyer, E. (1972). Permanent microscopic mounts. *Mycologia* **64,** 666–669.

Onions, A. H. S. (1971). Preservation of fungi. *Methods Microbiol.* **4,** 113–151.

Perkins, D. D. (1962). Preservation of *Neurospora* stock cultures with anhydrous silica gel. *Can. J. Microbiol.* **8,** 591–594.

Pollack, F. G. (1967). A simple method for preparing dried reference cultures. Mycologia 59: 541–544.

Tuite, J. (1969). "Plant Pathological Methods." Burgess, Minneapolis, Minnesota.

Models

INTRODUCTION

The use of models in tree pathology is a convenient method of reducing often complex phenomena to a simple diagram. The term "model" suggests that the diagram should be representative of a large number of closely related entities or events. Unfortunately, nature does not always produce events that can be conveniently expressed by a given model, and most models have exceptions. Some of the models that are useful in understanding or simplifying phenomena associated with tree diseases and the pathogens that cause them are presented in this chapter.

THE DISEASE CYCLE

For infectious disease to occur a certain sequence of events is required. The disease cycle (Fig. 6.1) is a model that represents the order and relationship of these events. A disease is initiated when inoculum is transferred to a portal of entry in the host. This is called inoculation. Inoculum may consist of any pathogen or viable propagules of the pathogen. The pathogen then penetrates the host tissue and becomes established by utilizing the host as a food source. The

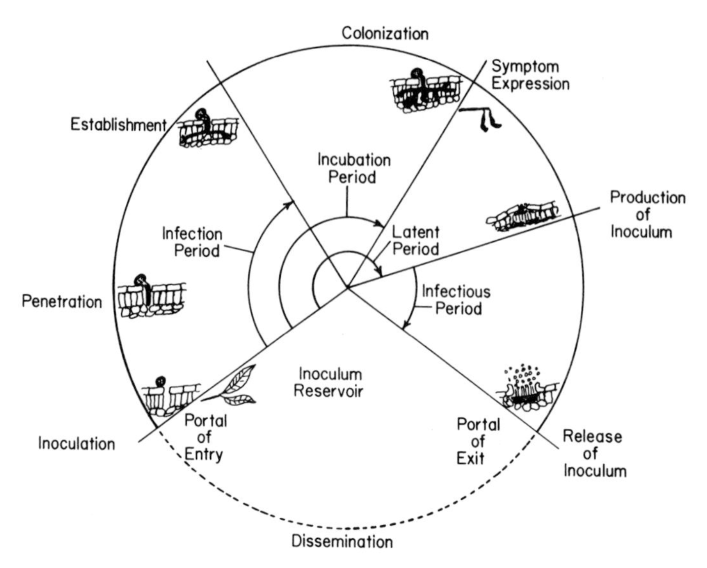

Fig. 6.1 Model of generalized disease cycle showing sequence of events in disease progression.

time required for the pathogen to enter the host and establish itself is called the *infection period.* The pathogen then proliferates and colonizes the host tissue, resulting in the appearance of symptoms. The time between effective inoculation and symptom expression is known as the *incubation period.* For the disease cycle to continue, the pathogen must produce new inoculum. The time from effective inoculation to the production of new inoculum is called the *latent period.* The period during which inoculum is being produced by the pathogen on the host is termed the *infectious period.* Inoculum leaves the host through a portal of exit and is either transported to a new host or returns to an inoculum reservoir. The inoculum reservoir may be an alternate host on which the pathogen lives as a parasite, residues of plant material on which the pathogen lives as a saprophyte, or a resistant resting state of the pathogen. If the disease cycle is interrupted at any point, the disease will be checked. Disease control is designed to focus on one or more of the events in the disease cycle.

THE DISEASE TRIANGLE

Factors that affect the development and destructiveness of disease are susceptibility of the host population, virulence of the pathogen, and an environment favorable to the pathogen. The effect of these three factors can be represented by a model termed the disease triangle (Fig. 6.2). If the maximum effect of each factor is represented by the three sides of an equilateral triangle, then the

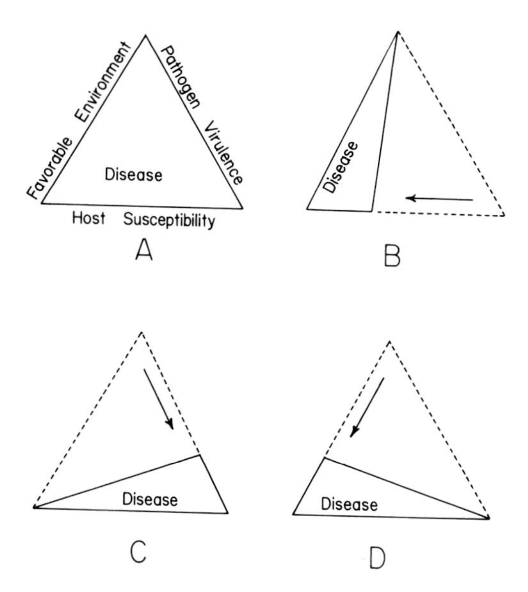

Fig. 6.2 Model of disease triangle showing factors involved in disease (A) and how severity of disease will decrease by reducing host susceptibility (B), pathogen virulence (C), or by changes to less favorable environmental factors (D).

maximum severity of the disease is represented by the area within the triangle (Fig. 6.2A). Reductions in host susceptibility (Fig. 6.2B), or in pathogen virulence (Fig. 6.2C), or changes to less favorable environmental factors for the pathogen (Fig. 6.2D), will reduce the severity of disease as represented by reductions in the areas of triangles B, C, and D, respectively, in Fig. 6.2. It is unlikely that any one factor will be at its maximum in any disease situation. Not represented are the possibilities that two or all three factors can be less than maximum at the same time. The model demonstrates that either complete host resistance, lack of pathogen virulence, or an environment in which the pathogen cannot grow or spread will eliminate disease.

THE CONTROL TRIANGLE

The model for a control program is standard for many programs or events that can be reduced to three components. Specifically this model demonstrates that three components are involved in the synthesis and application of a disease control program: biological, ecological, and economic (Fig. 6.3). It is not enough to kill the pathogen or to reduce its population. These involve only the *biological* aspect of a control program. The methods used to control the pathogen may be incompatible with other living things and the environment in which they live. The

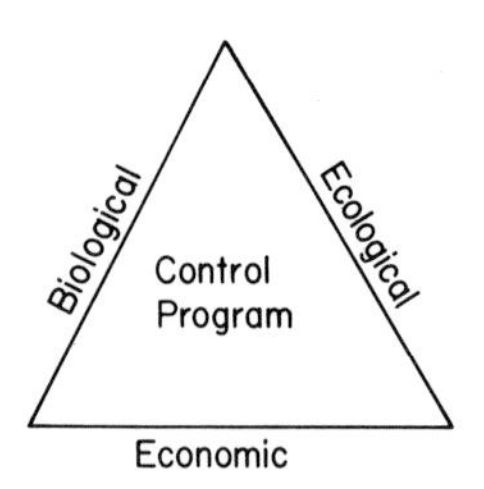

Fig. 6.3 Model of control triangle showing factors involved in overall control program decisions.

ecological implications must be weighed. Finally, it is important to recognize that control efforts are seldom applied when the cost of control is more than the value gained by their implementation. This involves *economics.*

ASSOCIATIONS AMONG ORGANISMS

The presence of disease usually signifies that a pathogen has parasitized a host. This is disease in its simplest form. In nature, however, no organism functions autonomously and there are many associations among organisms which may or may not influence disease development. In its relationship with another organism, an organism may be harmful (−), may be beneficial (+), or may have no effect (0). Most simply, then, there are six potential associations between two organisms. Several authors have variously named these associations, expanded on them, and defined them. For our discussion we will use terms that have common usage and will not dwell on all the subtle and detailed variations which undoubtedly exist. The six associations are

0/0 = Neutralism. Neither organism has an effect on the other.
+/0 = Commensalism. One organism is benefited without affecting the other.
−/0 = Amensalism. One organism is harmed without affecting the other.
+/− = Parasitism. One organism is benefited at the expense of the other.
+/+ = Symbiosis. Both organisms are benefited.
−/− = Reciprocal antagonism. Each organism adversely affects the other.

Complex diseases are often easier to visualize if models are made representing the associations of the organisms involved. For example, an associations model for beech bark disease might look like Fig. 6.4. Each arrow points away from the organism generating the effect illustrated by the symbol next to the arrow.

Beech is infested by the scale insect *Cryptococcus fagisuga,* which penetrates the bark with its stylet. This opening is supposedly the entry point for the parasitic fungus *Nectria coccinea* var. *faginata.* The fungus has no effect on the insect, but the insect has been beneficial to the fungus by producing a portal of entry. The twice-stabbed lady beetle feeds on the scale insect, and the fungus *Gonatorrhodiella highlei* is parasitic on *N. coccinea* var. *faginata.* Finally, the fungus *As-*

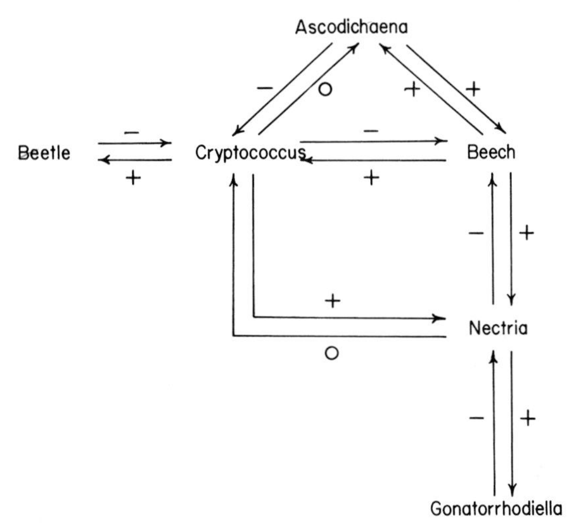

Fig. 6.4 Model of associations of organisms involved in beech bark disease where the relationship of one organism to another is harmful (−), beneficial (+), or has no effect (0).

codichaena rugosa in some way excludes the scale insect from areas of bark colonized by the fungus.

Selected References

Burkholder, P. R. (1952). Cooperation and conflict among primitive organisms. *Am. Sci.* **40,** 601–631.

Dindal, D. L. (1975). Symbiosis: Nomenclature and proposed classification. *Biologist (Phi Sigma Soc.)* **57,** 129–142.

7

Symptomatology

INTRODUCTION

Symptomatology is the study of disease symptoms, a symptom being defined as any condition resulting from disease that indicates its occurrence. We generally restrict the definition of symptom to mean a readily apparent morphological change, recognizing that changes do occur at the cellular level before they become visible. Frequently, a distinction is made between symptoms (Fig. 7.1) and signs (Fig. 7.2) of disease. A symptom of disease is expressed as a reaction of the host to a causal agent, whereas, a sign is evidence of disease other than that expressed by the host. Signs are ususally structures of pathogens. A disease is first noticed by the presence of symptoms and/or signs, and recognition of the specific type of symptom or sign will aid in the eventual diagnosis of the disease.

SYMPTOMS

Symptoms are divided into three general categories:

(1) necrotic symptoms—those symptoms that result from cessation of function leading to death;

Fig. 7.1 Illustrative drawings of various symptoms of disease.

(2) hypoplastic symptoms—those symptoms that result from underdevelopment or retardation of function;
(3) hyperplastic symptoms—those symptoms that result from overdevelopment or acceleration of function.

Necrotic Symptoms

1. Blight—Rapid killing of foliage, blossoms, and twigs.
2. Blotch—Large, irregular lesions on leaves, shoots, and stems.
3. Canker—Necrotic, often sunken lesions in the cortical tissues of stems and roots.

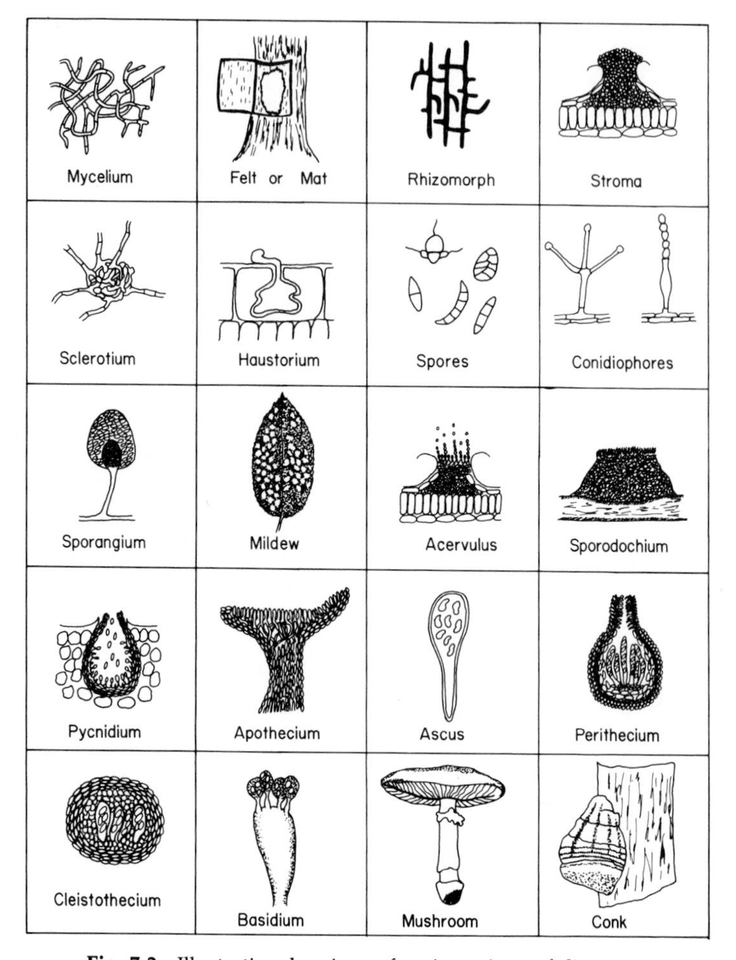

Fig. 7.2 Illustrative drawings of various signs of disease.

4. Decay—Disintegration of dead tissues.
5. Dieback—Progressive death of twigs and branches from their tips toward the trunk.
6. Hydrosis—A water-soaked, translucent condition of the tissue due to cell sap passing into intercellular spaces.
7. Scald—Blanching of the epidermis and adjacent tissues.
8. Scorch—Browning of leaf margins resulting from death of tissues.
9. Shot-hole—Circular holes in leaves resulting from the dropping out of the central necrotic areas of spots.
10. Spot—Lesions, usually defined, circular or oval in shape, with a central necrotic area surrounded by variously colored zones.

11. Wilt—Leaves or shoots lose their turgidity and droop.
12. Yellowing—Leaves turn yellow due to a degeneration of chlorophyll.

Hypoplastic Symptoms

1. Chlorosis—Failure of chlorophyll development in normally green tissues.
2. Dwarfing—Subnormal size in an entire plant or some of its parts.
3. Etiolation—Yellowing due to lack of light.
4. Rosetting—Crowded condition of foliage due to lack of internode elongation.
5. Suppression—Prevention of the development of certain organs.

Hyperplastic Symptoms

1. Anthocyanescence—Purplish or reddish coloration of leaves or other organs due to overdevelopment of anthocyanin pigment.
2. Callus—Overgrowth of tissues at the margins of wounds and diseased tissues.
3. Curl—Rolling or folding of leaves due to localized overgrowths of tissue.
4. Fasciation—Flattened condition of a plant part that is normally round.
5. Fasciculation or Witches' broom—Broomlike growth of densely clustered branches.
6. Heterotopy—Development of more or less normal tissues or organs in the wrong place.
7. Sarcody—Abnormal swelling of tissues above girdled branches or stems.
8. Scab—Roughened, crustlike lesion.
9. Tumefaction—Tumorlike or gall-like overgrowth of tissue.
10. Virescence—Development of chlorophyll in tissues where it is normally absent.

SIGNS

Signs are divided into three general categories: (1) vegetative structures—those pathogen structures that function primarily in absorption and storage of nutrients; (2) reproductive structures—those pathogen structures that function in reproduction of the organism; (3) disease products—gases and exudation products resulting from disease.

Vegetative Structures

1. Felt—Densely woven mat of mycelium.
2. Haustorium—Absorbing organ of a fungus which penetrates a host cell without penetrating the plasma membrane.
3. Mycelium—A mass of fungal threads or hyphae.

4. Pathogen cells—Generally, masses of bacterial cells.
5. Plasmodium—Naked mass of protoplasm.
6. Rhizomorph—Cordlike strand of fungal hyphae.
7. Sclerotium—A hard, compact, resting body composed of fungal hyphae.

Reproductive Structures

1. Acervulus—Mat of hyphae, generally associated with a host, forming erumpent lesions with short, densely packed conidiophores.
2. Apothecium—Open, cuplike, ascus-containing fruiting body.
3. Ascus—Saclike structure containing ascospores formed as a result of karyogamy and meiosis.
4. Basidium—Characteristically club-shaped structure on which basidiospores are produced as a result of karyogamy and meiosis.
5. Cleistothecium—Closed ascus-containing fruiting body.
6. Conidiophore—Specialized hyphal branch on which conidia are produced.
7. Conk—Woody shelflike structure characteristic of many wood-rotting fungi.
8. Mildew—Cobwebby or powdery superficial growth usually on leaves.
9. Mold—Woolly or furry surface growth of mycelium.
10. Mushroom (=Toadstool)—Umbrella-shaped fruiting structure of many Basidiomycetes.
11. Perithecium—Characteristically flask-shaped, ascus-containing fruiting body with a small opening (ostiole) and a wall of its own.
12. Pseudothecium—Fruiting body bearing asci in locules within a stroma.
13. Pycnidium—Asexual, hollow fruiting body containing conidia.
14. Seed-bearing plants—Higher plants that parasitize trees.
15. Sorus—Mass or cluster of spores borne on short stalks.
16. Sporangium—Enlarged tip of specialized hyphal branch in which sporangiospores are produced.
17. Spore—General name for a single to several celled propagative unit in the fungi and other lower plants. Examples of spores with specific names are conidia, ascospores, basidiospores, zoospores, oospores, sporangiospores, aeciospores, urediospores, chlamydospores, and teliospores.
18. Sporodochium—Cushion-shaped stroma covered with conidiophores.
19. Stroma—Compact mass of fungal hyphae on or within which fruiting structures are formed.
20. Worms—Generally nematodes which are microscopic, wormlike animals that can cause disease.

Disease Products

1. Odor—Characteristic smell associated with some host-pathogen interractions.
2. Ooze—Viscid mass made up of plant juices and often pathogen cells.

Selected References

Buckland, D. C., Redmond, D. R., and Pomerleau, R. (1957). Definition of terms in forest and shade tree diseases. *Can. J. Bot.* **35,** 675–679.

Ehrlich, J. (1941). Etiological terminology. *Chron. Bot.* **6,** 248–249.

Hepting, G. H., and Cowling, E. B. (1977). Forest pathology: Unique features and prospects. *Annu. Rev. Phytopathol.* **12,** 431–450.

Stevens, F. L., and Young, P. A. (1927). On the use of the terms saprophyte and parasite. *Phytopathology* **17,** 409–411.

PART II

Infectious Diseases

Infectious diseases are caused by biotic (living) pathogens. These include microscopic organisms such as fungi, bacteria, viruses, mycoplasmalike organisms, and nematodes, as well as macroscopic organisms such as seed plants. The term "infectious" implies that one organism (the pathogen) has established a food relationship with another organism (the host) and is capable of producing propagules that can infect other hosts.

Signs of the infecting pathogen and symptoms expressed by the host are often characteristic and diagnostic for a particular infectious disease. However, some diseases are difficult to diagnose. For example, wilt symptoms caused by invasion of pathogens into the vascular system of stems may be indistinguishable from wilt symptoms caused by invasion of pathogens into roots. Second, leaf symptoms are not always caused by pathogens localized in the leaves. Finally, symptoms of some noninfectious diseases may be indistinguishable from symptoms caused by infectious agents. For these reasons, diagnosis often requires thorough examination of the whole tree and laboratory isolation and/or identification of the causal agent.

Most infectious diseases are caused by fungi, and the number of fungal diseases described in this section compared to diseases caused by other organisms reflects this fact. Therefore, we have grouped fungal diseases into chapters based on general symptoms or parts of host affected. These chapters include leaf diseases, wilt diseases, canker diseases, rust diseases, root diseases, and discoloration and decay. Diseases caused by other biotic pathogens are described in chapters labeled with the category or organisms treated. These include bacteria, viruses, mycoplasmalike organisms, nematodes, and seed plants.

8

Fungi

INTRODUCTION

In the taxonomic scheme of living things the organisms referred to collectively as Fungi have been variously placed. Traditionally they were included in the kingdom Plantae, but recently most mycologists agree that they should be placed in a kingdom of their own: Myceteae. However, their significance to mankind should not be underrated, regardless of where we put them.

The fungi are so abundant on the earth that they must be considered one of the more successful forms of life. They can live on a variety of substrates including living animal and plant tissues, wood, paper, ink, fabrics, insect carcasses, twine, glue, paint, electrical insulation, leather, food products, jet fuel, and the wax in our ears. While the fungi have evolved primarily as saprophytes, they have developed various types of parasitism and are the most common and most important causes of disease in trees.

The basic unit of a fungus is the cell including cell wall, plasma membrane, nucleus with nuclear envelope, and various organelles within cytoplasm. No chlorophyll is present making the fungi heterotrophic, that is, they must use organic compounds as the primary source of energy. The primary constituent found in the cell walls is chitin, but the total composition of cell walls in the various species of fungi is not consistent nor clearly understood.

Some fungi go through their entire life cycle as individual cells performing all functions, without differentiation into tissues. Typically, however, the *thallus* of a fungus is composed of tubular, branching filaments, called *hyphae* that may or may not have crosswalls (*septa*). Exactly when a singular *hypha* becomes a plural *hyphae* is arbitrary. As several hyphae become intimately enmeshed, they collectively are called *mycelium*.

Reproduction in fungi is accomplished in many varied ways, each classified into one of two types: sexual and asexual (Fig. 8.1). The product of both types is some form of *spore*. Terms which are used interchangeably with sexual and asexual are *perfect* and *imperfect*, respectively.

SEXUAL REPRODUCTION

Sexual reproduction involves the union of two compatible nuclei. The total sexual process consists of three distinct phases, the length of which varies considerably. The first phase is *plasmogamy* in which the union of two protoplasts

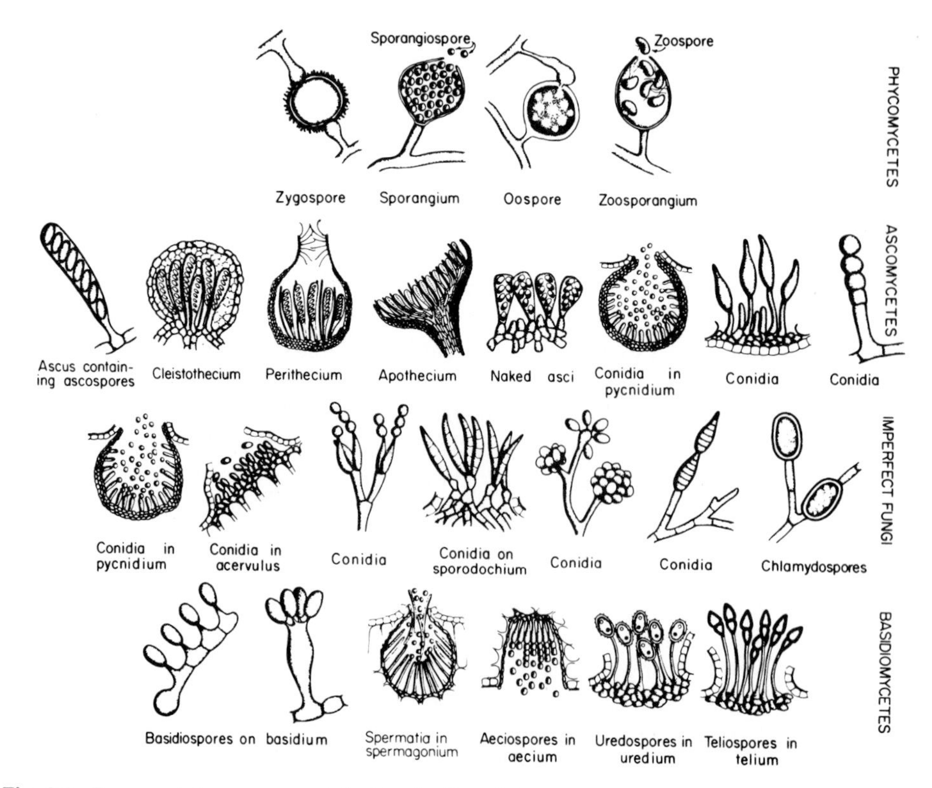

Fig. 8.1 Representative spores and fruiting bodies of the main groups of fungi. (Drawing courtesy of George N. Agrios, "Plant Pathology," 2nd ed. Academic Press, New York, 1978.)

brings two or more compatible nuclei into close association. Plasmogamy is followed by *karyogamy*, the actual fusion of two nuclei. Prior to fusion, the nuclei are in the haploid ($1n$) state; after fusion the combined nucleus is in the diploid ($2n$) state. The sexual process is completed when the diploid nucleus undergoes *meiosis*, forming four haploid nuclei. Depending on the species, the nuclei may then undergo one to several mitotic divisions.

ASEXUAL REPRODUCTION

Asexual reproduction involves the formation of propagules that have not resulted from the union of compatible nuclei. Generally, the asexual state is repeated many times during a season, whereas, the sexual state often occurs only once. From this standpoint the asexual state functions primarily in the propagation of the species by producing large numbers of individuals, whereas the sexual state functions primarily in the adaptive survival of the species by producing recombinations of individuals.

CLASSIFICATION

Although the phylogenetic origin of species of fungi may not be important in diagnoses of tree diseases caused by fungi, some understanding of their classification may be helpful. The better pathogenic organisms are understood, the better the chances of controlling them. Also, two pathogenic organisms that have been closely allied taxonomically may be similarly controlled. Of the more than 100,000 species of fungi, less than 2000 have been proved pathogenic on trees. Individuals have been grouped from the standpoint of morphology, nutrition, biochemistry, ontogeny, and other characteristics. The paucity of fossil records limits our understanding of the phylogenetic relationships, but the tree pathologist can function very well with the slow evolution of fungal taxonomy. To the pathologist, identification of a causative agent is paramount and this requires mostly an association between terminology and morphology. The question becomes: Can the organism be identified so that published research information can be utilized? It matters little that *Puccinia sparganioides* or *Puccinia peridermiospora* causes ash rust. Last year's organism and this year's name do not change control strategies. The fungus is still the same fungus.

Traditionally, the fungi are divided into four major groups: Phycomycetes, Ascomycetes, Basidiomycetes, and Deuteromycetes. Slime molds and cellular slime molds have been included with fungi, although they have characteristics which are common to both fungi and protozoa. Since the slime molds have not as yet been implicated in tree disease, they will not be discussed further.

1. Phycomycetes—generally referred to as the "Lower Fungi" because of their supposedly primitive characteristics. Hyphae are typically coenocytic (nonsep-

tate). Asexual reproduction is by spores (often motile) produced in sporangia. Morphological structures and mechanisms of sexual reproduction are varied, but the sexual process generally results in the production of a resistant resting spore. This large group is divided into several classes, but only a few major genera are tree pathogens.

2. Ascomycetes—generally referred to as the "Higher Fungi" along with Basidiomycetes and Deuteromycetes. Hyphae are septate. The distinguishing character is the *ascus,* a saclike structure containing ascospores formed as a result of karyogamy and meiosis. The asci arise from a mat of fertile hyphae, called a hymenium, and in most species are enclosed in some type of fruiting structure called an *ascocarp*:

a. cleistothecium—ascocarp completely closed, spherical.
b. perithecium—ascocarp with a small opening and a wall of its own, flask-shaped.
c. apothecium—ascocarp open, cup-shaped.
d. pseudothecium—ascocarp bearing asci in locules within a stroma.

Asexual reproduction is by production of conidia, borne on conidiophores that may occur singly, in groups, on mats of hyphae, or in enclosed fruiting structures called *pycnidia.* Ascomycetes include causal agents of a variety of diseases such as foliage diseases, wilts, and cankers.

3. Basidiomycetes—distinguishing character is the *basidium,* a structure with a function similar to the ascus. However, although *basidiospores* are the product of karyogamy and meiosis, as with ascospores, they are borne on the surface of the basidium. Hyphae are septate, and in many species exhibit a complicated septal apparatus called a dolipore septum. Other structures known as clamp connections are found on the hyphae of many species. Rusts and smuts do not produce a fruiting structure for the basidia, but most other Basidiomycetes do. These "basidiocarps" are commonly recognized as mushrooms, conks, puffballs, earthstars, stinkhorns, and bird's nests. Asexual reproduction is by budding, mycelial fragmentation, or production of conidia. Diseases most commonly attributed to Basidiomycetes include rust diseases and wood decay.

4. Deuteromycetes—characterized by the absence of a sexual state. Hyphae are septate. Reproduction is the same as the asexual state of Ascomycetes. Because of the similarity of the conidial state of these nonsexual fungi to the conidial state of many Ascomycetes, it is presumed that the Deuteromycetes are conidial states of Ascomycetes (and to a lesser extent Basidiomycetes) which have lost their sexual function. It is also distinctly possible that we have not found or associated Ascomycetes with known Deuteromycetes.

Selected References

Alexopoulos, C. J., and Mims, C. W. (1979). "Introductory Mycology," 3rd ed. Wiley, New York.
Barnett, H. L., and Hunter, B. B. (1972). "Illustrated Genera of Imperfect Fungi." Burgess, Minneapolis, Minnesota.

Leaf Diseases

INTRODUCTION

Leaf diseases occur on most species of trees and shrubs. A variety of symptoms can occur (see Fig. 7.1):

1. *Leaf spot* is characterized by usually discrete lesions that are at first discolored and then may become necrotic, turning brown or black as tissues die. Sometimes the dead tissues fall out, causing a "shot-hole" appearance. On needlelike leaves the tissues above the lesion often die for lack of water.
2. *Leaf blotch* is similar to leaf spot except that the lesions tend to be larger, somewhat irregular, and not as clearly delimited.
3. *Anthracnose* is even more extensive than blotch, found on leaf margins, across and along veins, and often over the entire leaf.
4. *Mildew* is more sign than symptom. It is caused by fungi that penetrate epidermal cells, but the cells are not killed. As the fungi spread over the leaf surfaces, they produce spores and mycelia that appear white and powdery.
5. *Leaf cast* is most prevalent on conifers (needle cast) and is a collective symptom given to diseases that result in the death and shedding of leaves.
6. *Leaf blight* is characterized by general and rapid killing of the leaf.

Other symptom names have been coined (e.g., "scab," "blister"), but the preceding six, or combinations of them, will probably describe most leaf disease symptoms.

Leaf diseases can be caused by many agents, but fungi are responsible for most. Fruiting structures of the pathogens can often be seen within the limits of the lesions. Although there are a few serious leaf diseases, most cause little harm to the tree and are more of an aesthetic problem than a pathological one. Therefore, forest trees outside of the nursery are seldom treated, but increased awareness of problems with shade trees has prompted the use of control strategies in the urban environment. It is important to realize, however, that once a tree is infected, particularly a deciduous tree, control is seldom feasible until the following year. Some consolation can be found in the fact that a broad-leaved deciduous tree suffering complete defoliation from a leaf pathogen probably will refoliate the following spring with little evidence of harm. Repeated years of defoliations, however, and efforts to refoliate the same season, can lead to death of the tree by reducing its vigor and increasing its susceptibility to other more serious pathogens.

Most leaf pathogens enter the leaves in the spring when cool moist weather is prevalent. Because deciduous trees lose all their leaves at the end of each growing season, whereas evergreens lose only the oldest of approximately 3 years'

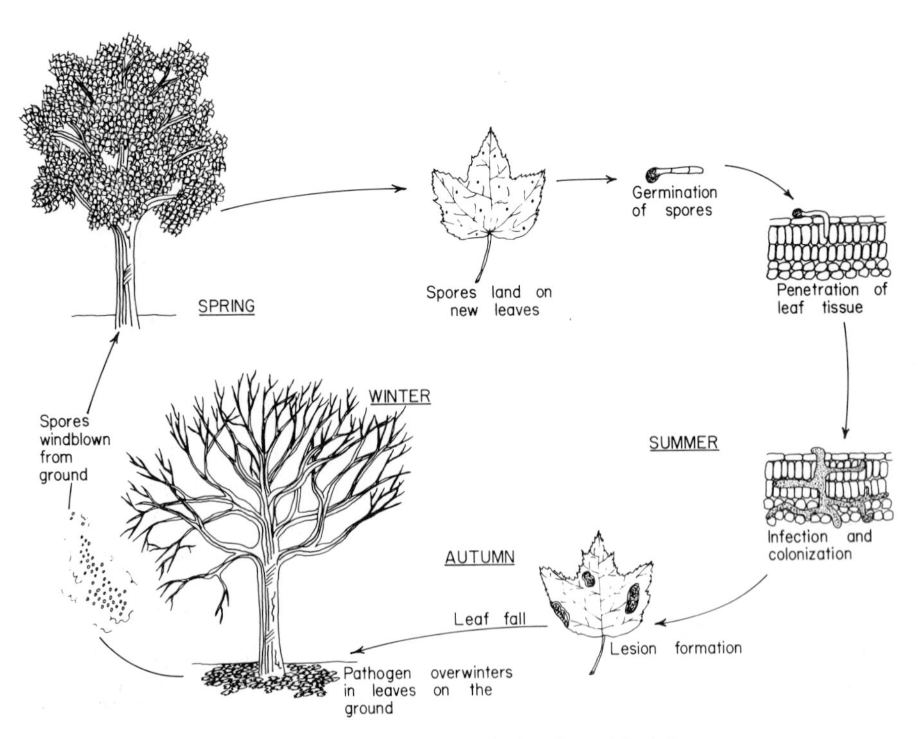

Fig. 9.1 Typical disease cycle of a hardwood leaf disease.

worth of leaves, the generalized disease cycles differ somewhat between the two groups.

Generally, inoculation, germination, penetration, infection, and colonization of host tissues by a pathogen are the same for both deciduous and evergreen trees (Figs. 9.1, 9.2). However, many pathogens that infect leaves of deciduous trees overwinter in leaves that have fallen to the ground. Generally, pathogens that infect new leaves of evergreen trees remain in contact with the host for at least 1–2 years. Older leaves that have become infected will carry the pathogen to the ground during normal leaf fall. *Where* the pathogen overwinters may be important when considering recommendations for control.

It is often difficult to discriminate between diseases that involve only the leaves and those that are merely manifestations of other problems. For example, diseases of roots and stems are often witnessed first as leaf symptoms. Careful examination of the whole tree may be necessary for correct diagnosis.

DISEASE: POWDERY MILDEW

Primary causal agents: Species of *Phyllactinia, Erysiphe, Microsphaera, Uncinula, Podosphaera,* and *Sphaerotheca*

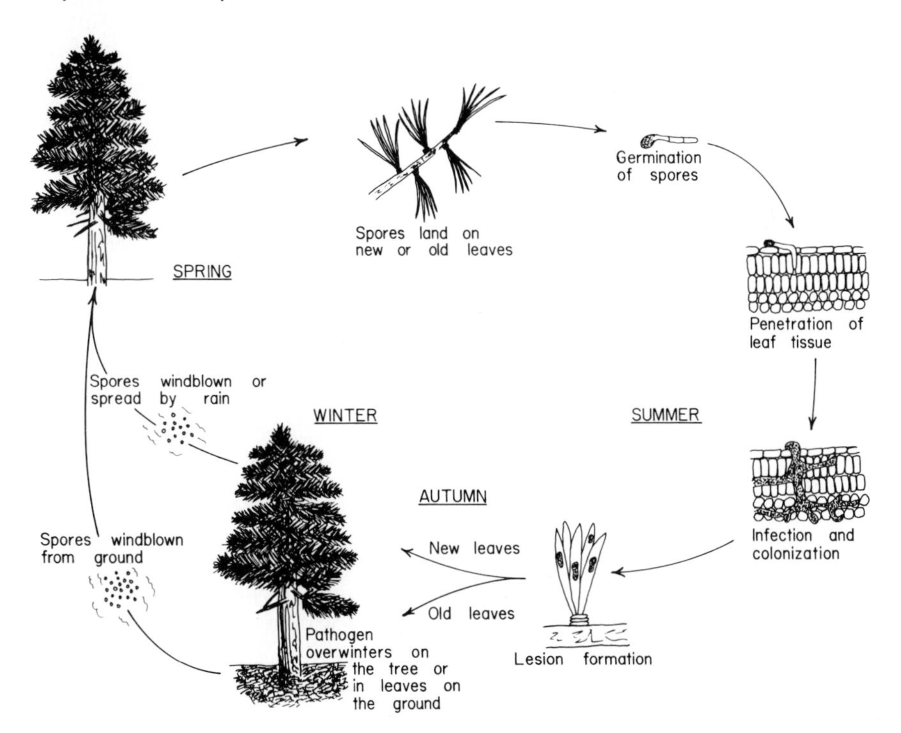

Fig. 9.2 Typical disease cycle of an evergreen leaf disease.

Fig. 9.3 Powdery mildew on oak (A) and lilac (B).

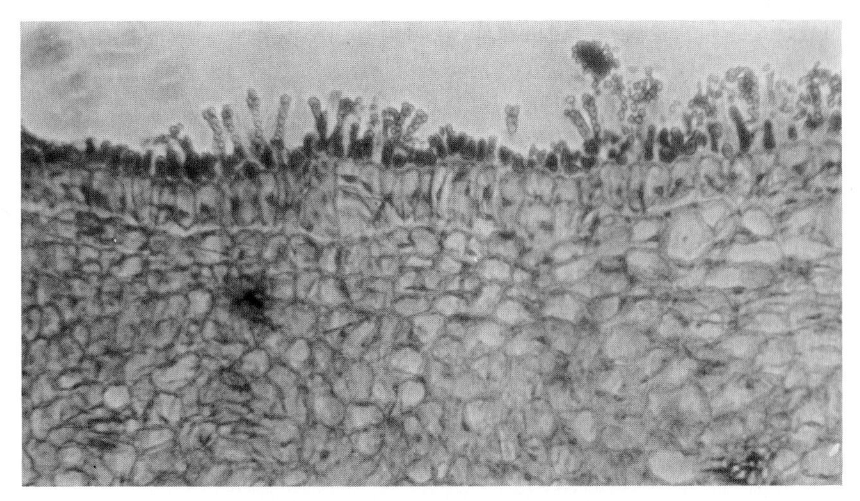

Fig. 9.4 Cross section of a leaf with mycelium, conidiophores, and conidia of a powdery mildew fungus on the surface. 250×. (Photographs from microscope slides prepared by Triarch, Inc., Ripon, Wisconsin.)

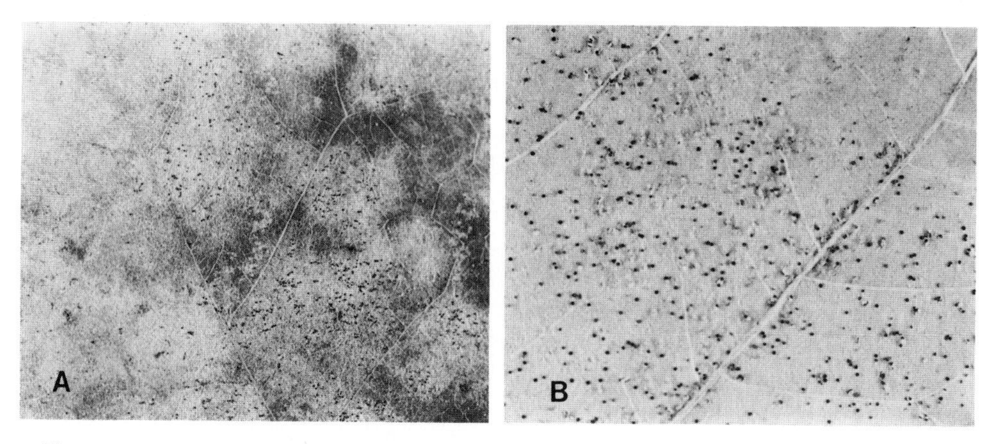

Fig. 9.5 Mycelium and cleistothecia of a powdery mildew fungus on leaf surface (A), closeup (B).

Hosts: Most broad-leaved trees and shrubs as well as many herbaceous plants

Symptomatology: From mid to late summer, infected leaves look as though dusted with powder (Fig. 9.3), hence the name "powdery mildew." The symptoms have sometimes been mistaken for road dust on hosts such as oaks

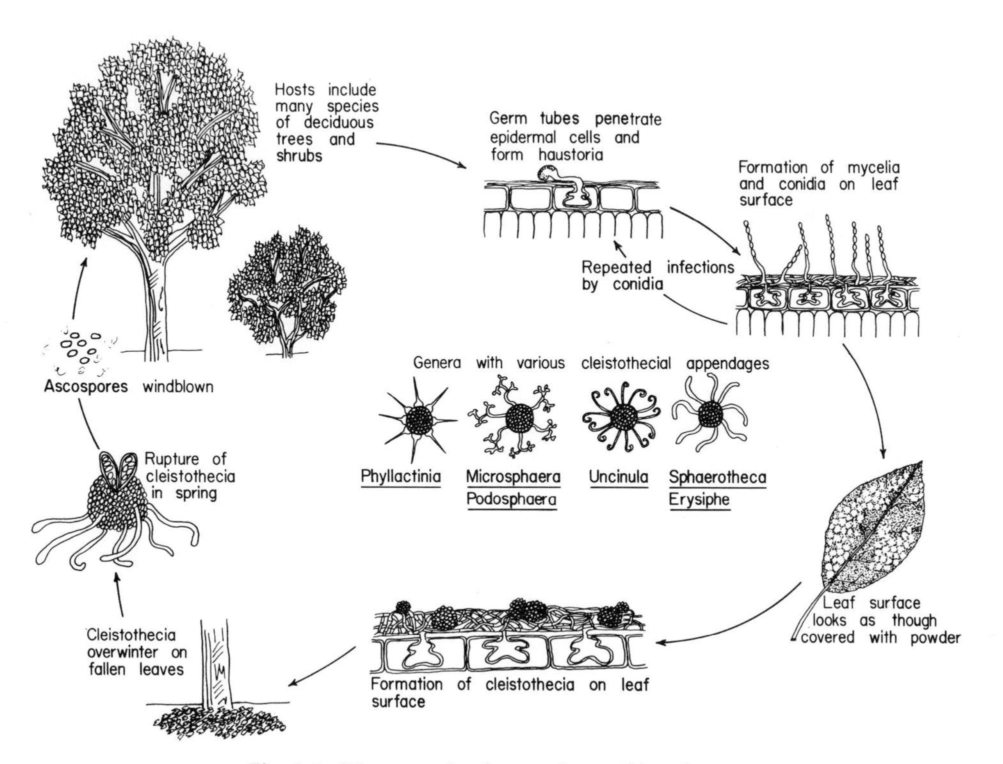

Fig. 9.6 Disease cycle of a powdery mildew fungus.

and lilacs planted near dirt roads. Actually, this "dust" or "powder" is the pathogen itself and is a combination of mycelium and asexual spores growing on the leaf surfaces (Fig. 9.4). Some chlorosis may be seen in tissues beneath the fungus. During the fall, tiny specks, first orange and later black, can be seen throughout the powdery material (Fig. 9.5). These are the sexual fruiting

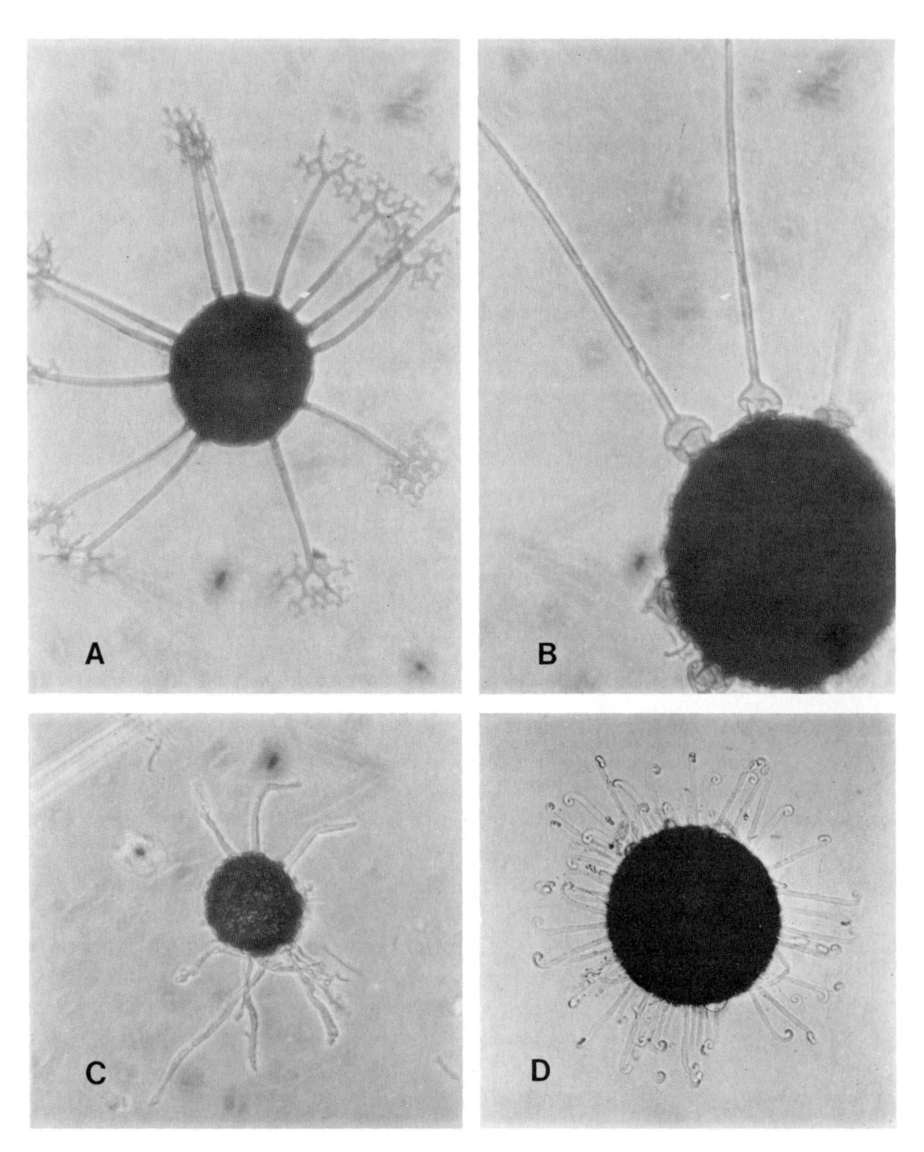

Fig. 9.7 Cleistothecia with appendages. Dichotomously branched (A), bulbous base (B), mycelioid (C), coiled tips (D). (Photographs from microscope slides prepared by Triarch, Inc., Ripon, Wisconsin.)

structures (cleistothecia) of the fungus. Occasionally, symptoms such as discoloration, dwarfing, or distortion may be seen on heavily infected leaves, especially those that were infected early in the season.

Etiology: Germ tubes of the pathogen penetrate walls of the leaf epidermal cells (Fig. 9.6). The plasma membranes are not penetrated, but the fungus forms absorbing structures known as haustoria. The fungus proliferates over the leaf surface, obtaining nourishment from haustoria in the epidermal cells. Hyaline, barrel-shaped conidia (Form genus = *Oidium*) are produced in large numbers during the summer and are wind blown to other susceptible hosts, initiating new infections. Cleistothecia form during late summer and fall, and all have characteristic appendages that help to identify them (Fig. 9.7). A simple key to the genera of powdery mildews is as follows:

A. One ascus per cleistothecium (Fig. 9.8B)
  B. Appendages mycelioid .... **Sphaerotheca**
  BB. Appendages dichotomously branched .... **Podosphaera**
AA. Several asci per cleistothecium (Fig. 9.8A)
  C. Appendages mycelioid .... **Erysiphe**
  CC. Appendages not mycelioid
    D. Appendages dichotomously branched .... **Microsphaera**
    DD. Appendages not dichotomously branched
      E. Appendages with a bulbous base .... **Phyllactinia**
      EE. Appendages with coiled tips .... **Uncinula**

The cleistothecia overwinter on fallen leaves. The following spring, the cleistothecia break open (Fig. 9.9). Asci release ascospores which are wind blown to susceptible hosts.

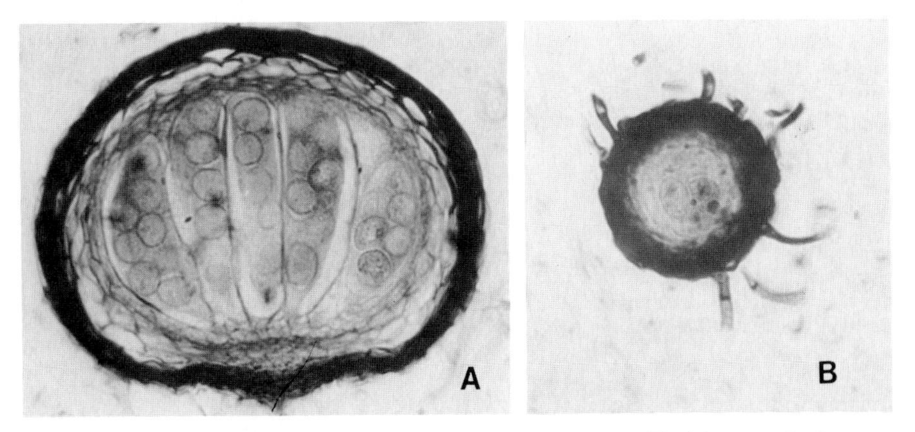

Fig. 9.8 Cleistothecia containing several asci (A) and only one ascus (B). (Photographs from microscope slides prepared by Triarch, Inc., Ripon, Wisconsin.)

Fig. 9.9 Ruptured cleistothecium of *Uncinula* showing asci and ascospores. (Photograph from microscope slides prepared by Triarch, Inc., Ripon, Wisconsin.)

Control: Since damage is usually slight, control is seldom warranted. Surface applications of certain fungicides control the disease. Destroying fallen leaves may help by limiting potential inoculum.

Selected References

Harris, J. L., and Roth, I. L. (1974). Scanning electron microscopy of perithecial development in a species of Phyllactinia on oak. *Can. J. Bot.* **52,** 2175–2179.

Harris, J. L., and Roth, I. L. (1975). Scanning electron microscopy of perithecial development in a species of Microsphaera on oak. *Can. J. Bot.* **53,** 279–283.

Massey, L. M. (1948). Understanding powdery mildew. *Am. Rose Annu.* pp. 136–145.

Yarwood, C. E. (1957). Powdery mildews. *Bot. Rev.* **23,** 235–301.

DISEASE: APPLE SCAB

Primary causal agent: *Venturia inaequalis* (Cke.) Wint.

Hosts: Apple (*Malus* spp.)

Symptomatology: Circular to irregular, olive-green spots appear on upper leaf surfaces in the spring (Fig. 9.10). The spots eventually become black and some coalesce. Heavy infections will cause leaf distortion and premature leaf fall. Infections on developing fruit appear as distinct scablike lesions that become somewhat raised and cracked (Fig. 9.10).

Fig. 9.10 Apple scab lesions on leaf (left) and fruit (right).

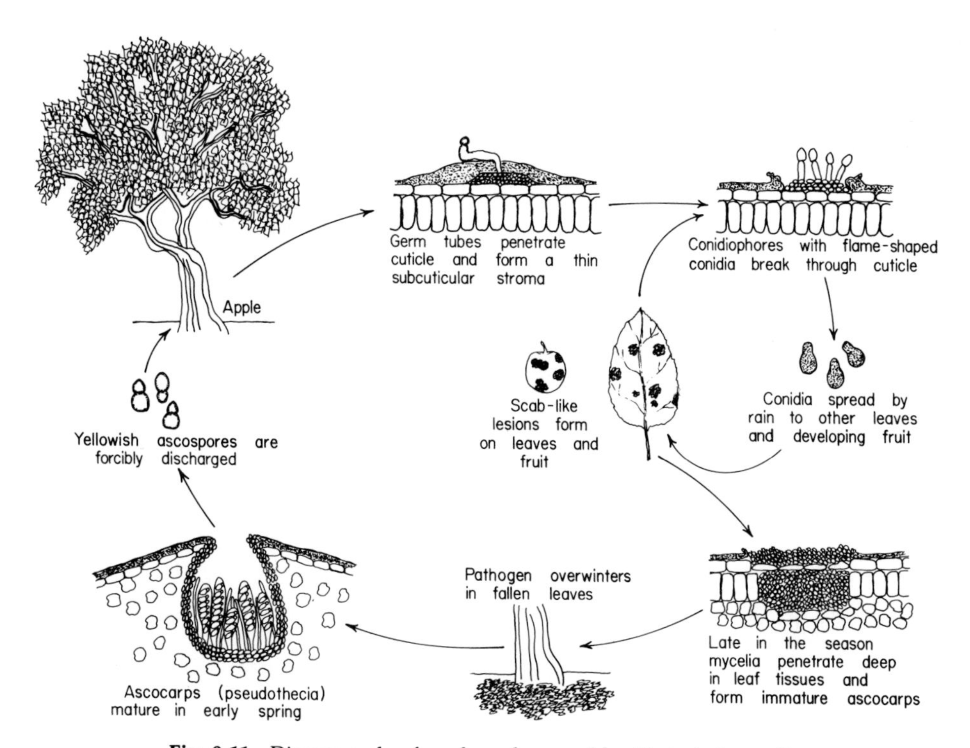

Fig. 9.11 Disease cycle of apple scab caused by *Venturia inaequalis*.

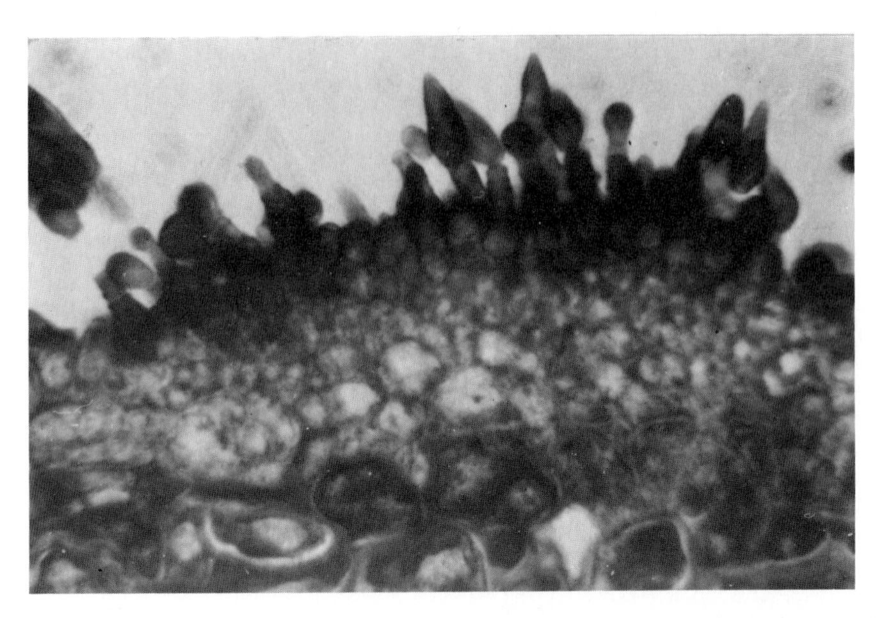

Fig. 9.12 Conidiophores and flame-shaped conidia arising from stroma of *V. inaequalis* on epidermis of apple leaf. (Photograph from microscope slides prepared by Carolina Biological Supply Company, Burlington, North Carolina.)

Etiology: Dark, unequally bicellular ascospores germinate in the spring and penetrate the cuticle of young leaves and developing fruit (Fig. 9.11). A thin stroma develops between the cuticle and epidermis. Conidiophores develop and produce large numbers of flame-shaped dark conidia which rupture the cuticle (Fig. 9.12). The conidia are washed or windblown to other leaves and fruit, causing secondary infections. Late in the growing season, mycelia penetrate deep in the infected leaf tissues and form stromatic, immature ascocarps which overwinter in the fallen leaves. The following spring the ascocarps (pseudothecia) mature and release yellowish ascospores (Fig. 9.13). The ascospores turn brown with age.

Control: Several protectant fungicides are available which provide excellent control of apple scab. However, because of repeated secondary cycles of the disease throughout the growing season, timely sprays associated with rainfall are required.

Selected References

Alexander, S. A., and Lewis, F. H. (1975). Reduction of apple scab inoculum with fungicides. *Plant Dis. Rep.* **59,** 890–894.

Jones, A. L., and Gilpatrick, J. D., eds. (1978). Proc. Apple and Pear Scab Workshop. *N.Y. Agric. Exp. Stn., Spec. Rep.* No. 28.

Julien, J. B. (1958). Cytological studies of *Venturia inaequalis*. *Can. J. Bot.* **36,** 607–613.

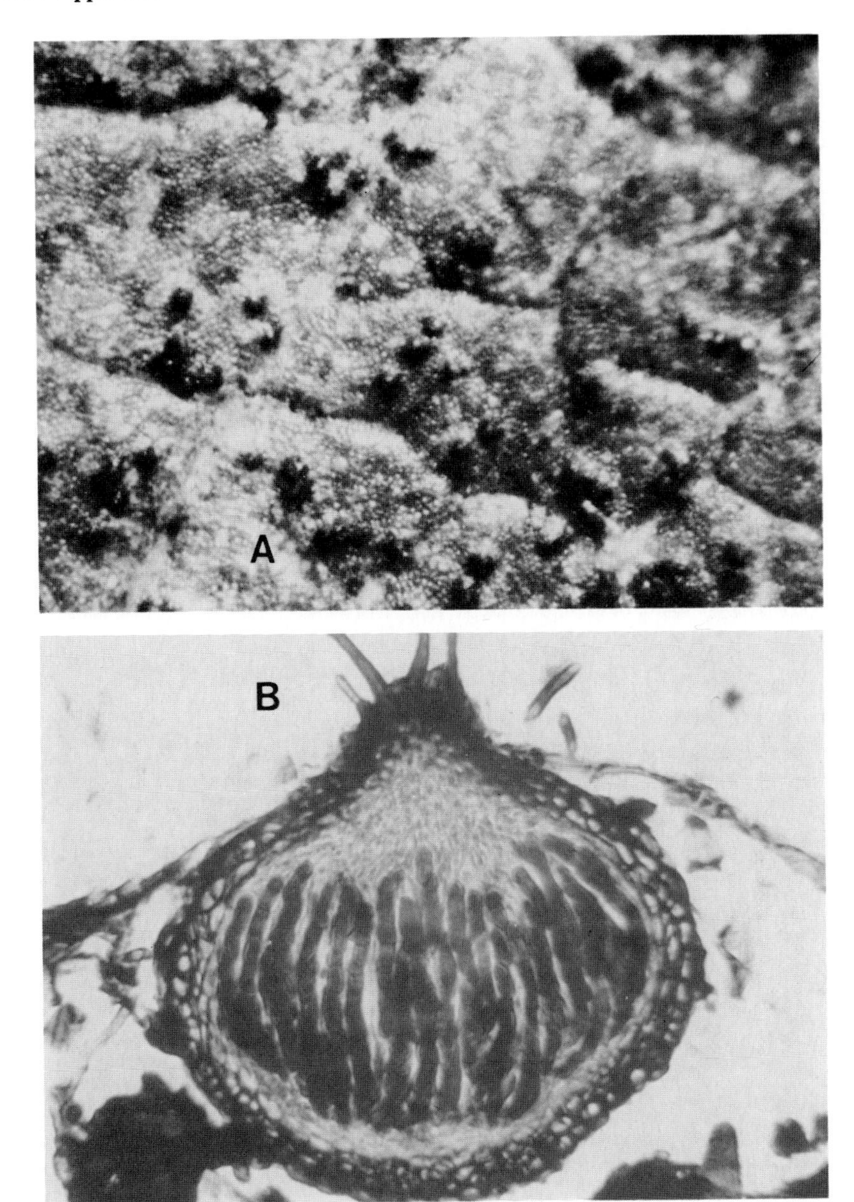

Fig. 9.13 Pseudothecia of *V. inaequalis* protruding through apple leaf surface (A) and cross section of a mature pseudothecium (B). (Photograph A courtesy of David Gadoury, University of New Hampshire, Durham. Photograph B from microscope slide prepared by Carolina Biological Supply Company, Burlington, North Carolina.)

Fig. 9.14 Anthracnose on leaves of elm (A), oak (B), and American sycamore (C). (Photographs A and B courtesy of George N. Agrios. Photograph C courtesy of Shade Tree Laboratories, University of Massachusetts, Amherst.)

MacHardy, W. E. (1979). A simple, quick technique for determining apple scab infection periods. *Plant Dis. Rep.* **63,** 199–204.

Ross, R. G., and Newbery, R. J. (1975). Effects of seasonal fungicide sprays on perithecium formation and ascospore production in *Venturia inaequalis*. *Can. J. Plant Sci.* **55,** 737–742.

Sutton, T. B., Jones, A. L., and Nelson, L. A. (1976). Factors affecting dispersal of conidia of the apple scab fungus. *Phytopathology* **66,** 1313–1317.

Fig. 9.15 Bud and twig mortality of American sycamore caused by *Gnomonia* spp. (Photograph courtesy of Shade Tree Laboratories, University of Massachusetts, Amherst.)

Fig. 9.16 Dieback and sprouting associated with cankers on American sycamore caused by *Gnomonia* spp. (Photograph courtesy of Shade Tree Laboratories, University of Massachusetts, Amherst.)

Fig. 9.17 Premature defoliation of American sycamore caused by *Gnomonia* spp. (Photograph courtesy of Shade Tree Laboratories, University of Massachusetts, Amherst.)

DISEASE: ANTHRACNOSE

Primary causal agents: *Gnomonia* spp.

Hosts: Wide variety of broad-leaved trees, but most severe on ash (*Fraxinus* spp.), maple (*Acer* spp.), oak (*Quercus* spp.), and sycamore (*Platanus* spp.)

Symptomatology: Depending on the host species, this disease may produce symptoms that are generally restricted to the leaves or may produce dieback and "canker-like" symptoms on shoots and branches. Symptoms on infected leaves vary from small necrotic spots to irregular lesions along leaf margins, across and along veins, and often over the entire leaf (Fig. 9.14). Infected buds and twigs are killed during early spring and appear as though frost-injured (Fig. 9.15). Branch cankers develop at the base of infected twigs (Fig. 9.16). Heavy infections often lead to premature defoliation (Fig. 9.17).

Etiology: Fungi that primarily infect leaves, overwinter in perithecia in fallen leaves (Fig. 9.18). During the spring, unequally bicellular ascospores are windblown to emerging leaves of the host tree, causing necrosis of infected tissues. Conidia are developed in the lesions and are rainsplashed to other leaves. Fungi that infect both stems and leaves overwinter in fruiting bodies

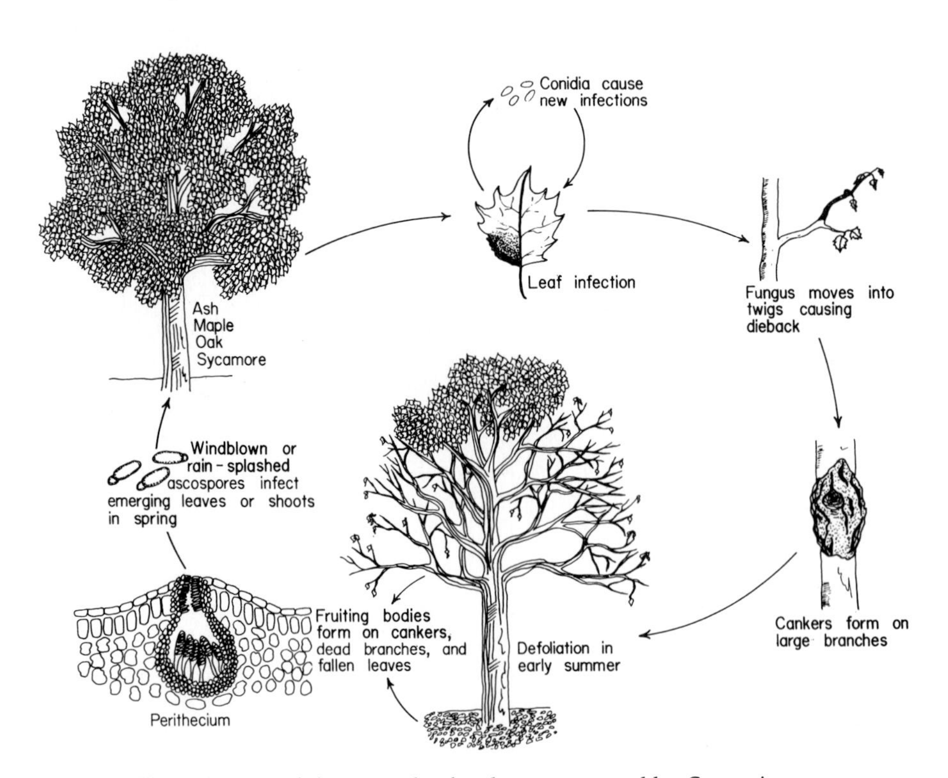

Fig. 9.18 Typical disease cycle of anthracnose caused by *Gnomonia* spp.

Fig. 9.19 Leaf spot of elm caused by *Gnomonia ulmea.*

on cankers, dead twigs, and fallen leaves, and in infected buds and twigs. The fungi spread into twigs from infected leaves, causing dieback the following spring. Further spread of the fungus can cause canker formation on larger branches. Active cankers may soon girdle and kill infected branches. An organism in the same genus, *Gnomonia ulmea* (Schw. ex Fr.) Thuem., causes leaf spot on elm (Fig. 9.19). Some pathologists consider this to be an anthracnose, while others treat it as a leaf spot.

Control: Destruction of plant materials in which the pathogen overwinters has been successful for some host species. This includes pruning infected twigs and branches and destroying fallen leaves. In severe cases some control can be obtained with several applications of protectant fungicides.

Selected References

Himelick, E. B. (1961). Sycamore anthracnose. *Proc. Int. Shade Tree Conf.* **37,** 136–143.
Neely, D. (1976). Sycamore anthracnose. *J. Arboricult.* **2,** 153–157.

Schuldt, P. H. (1955). Comparison of anthracnose fungi on oak, sycamore, and other trees. *Contrib. Boyce Thompson Inst.* **18,** 85–107.

Sinclair, W. A., and Johnson, W. T. (1968). Anthracnose diseases of trees and shrubs. *Cornell Tree Pest Leaf.* **A-2.**

DISEASE: OAK LEAF BLISTER

Primary causal agent: *Taphrina caerulescens* (Mont. & Desm.) Tul.

Hosts: Most species of oak (*Quercus* spp.)

Symptomatology: In early summer, distinct localized areas of tissue on the upper leaf surface turn light green and become raised, resembling blisters (Fig. 9.20). The raised blisters on the upper leaf surface are accompanied by sunken areas on the lower surface. Toward the end of the summer, the blistered tissues turn brown and die. Only very severe infections will cause defoliation.

Fig. 9.20 Raised blisters on oak leaf caused by *Taphrina caerulescens* (A), side view (B).

Etiology: Spores of the pathogen overwinter in bud scales (Fig. 9.21). During early spring the spores germinate and penetrate through stomata of the emerging leaves. Intercellular colonization causes hypertrophy of host tissues, and distinct blisters are formed on the upper leaf surfaces. The intercellular hyphae then produce naked asci (no fruiting structure) just beneath the cuticle. Pressure from the expanding asci soon ruptures the cuticle and the hyaline globose ascospores are released. The ascospores may bud, like yeasts, within the asci or after release. These budded spores (blastospores) are wind carried to bud scales of susceptible hosts.

Control: Control on forest trees is neither practical nor necessary. If control on shade trees is required because of rare heavy infection or for aesthetic reasons, one application of fungicide just before bud break will prevent disease occurrence.

Selected References

Kramer, C. L. (1960). Morphological development and nuclear behavior in the genus *Taphrina*. *Mycologia* **52,** 295–320.

Mix, A. J. (1949). A monograph of the genus *Taphrina*. *Univ. Kans. Sci. Bull.* **33,** Part. I(1), 3–167.

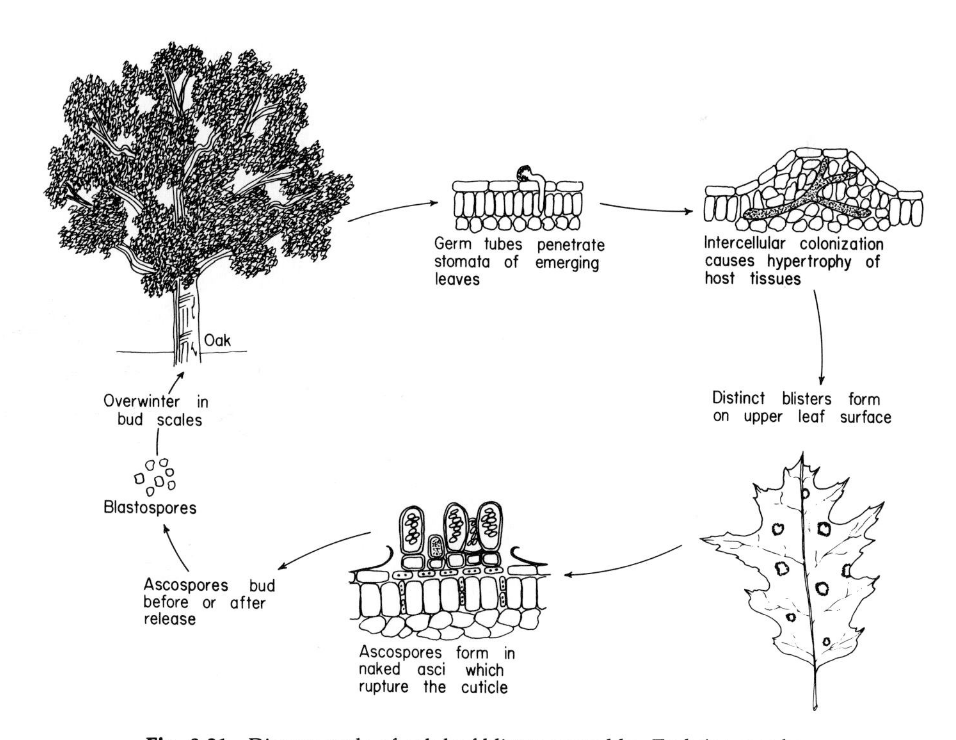

Fig. 9.21 Disease cycle of oak leaf blister caused by *Taphrina caerulescens*.

Fig. 9.22 Tar spot on silver maple caused by *Rhytisma acerinum* (A), closeup (B).

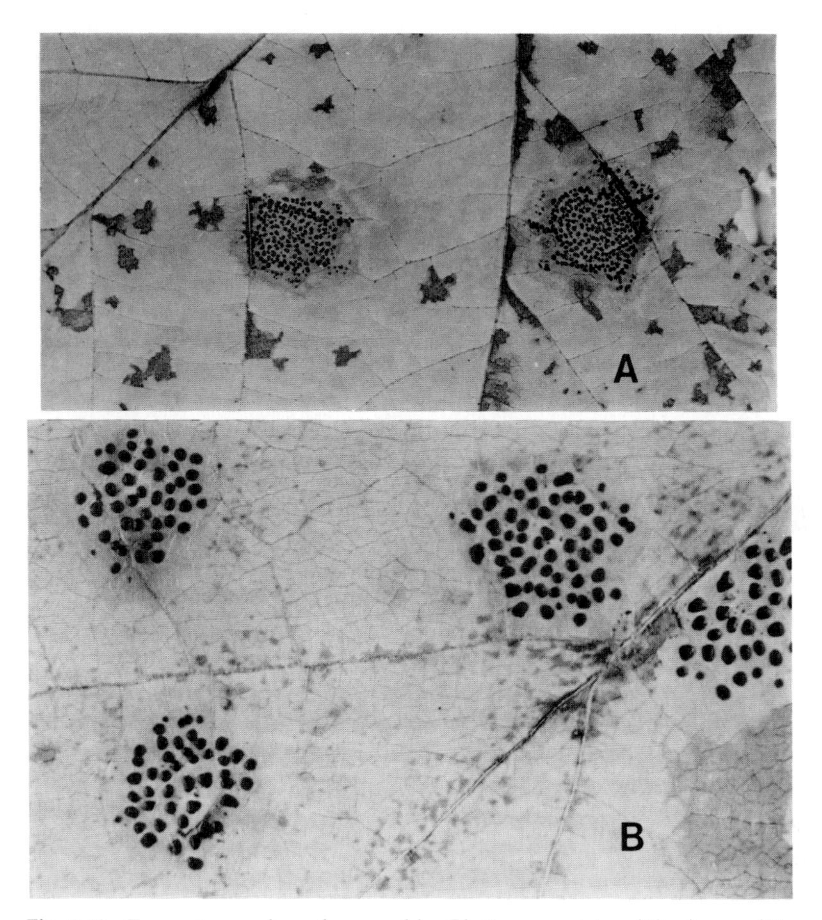

Fig. 9.23 Tar spot on red maple caused by *Rhytisma punctatum* (A), closeup (B).

DISEASE: TAR SPOT

Primary causal agents: *Rhytisma acerinum* Pers. ex Fr., *R. punctatum* Pers. ex Fr., *R. salicinum* Pers. ex Fr.

Hosts: *R. acerinum* and *R. punctatum* on several species of maple (*Acer* spp.) and *R. salicinum* on willows (*Salix* spp.)

Symptomatology: Soon after infection, leaves exhibit light yellow-green spots generally in groups. By midsummer the groups of individual small spots coalesce in *R. acerinum* (Fig. 9.22) and *R. salicinum* to form large, black, thickened stromata up to ½ inch (1–1.5 cm) across. The stromata resemble spots of tar, hence the name of the disease. In *R. punctatum,* however, the small

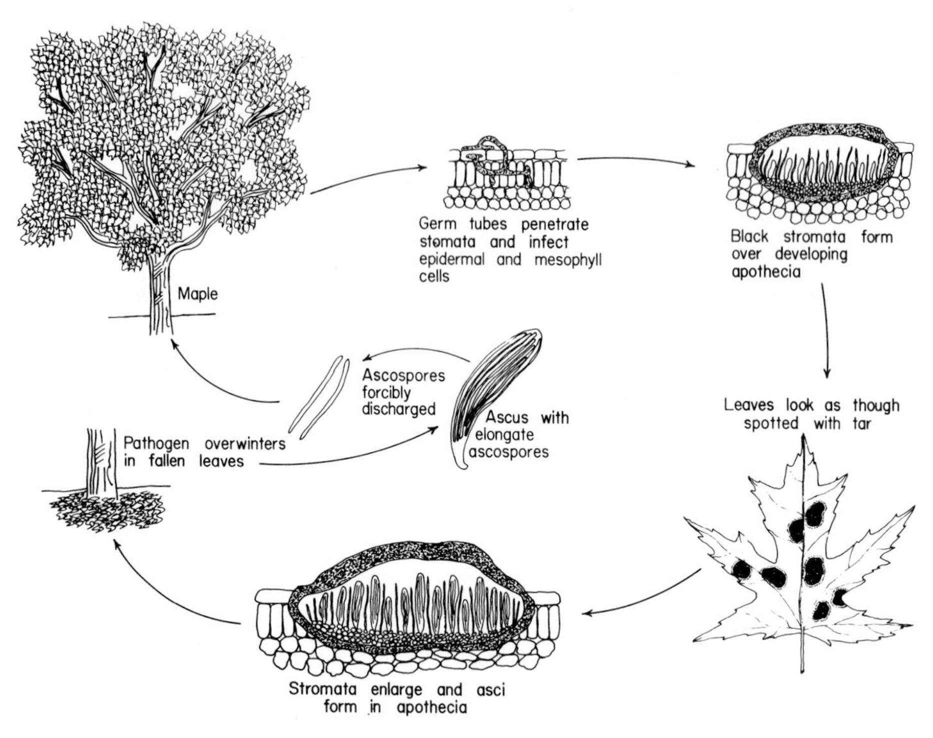

Fig. 9.24 Disease cycle of tar spot caused by *Rhytisma* spp.

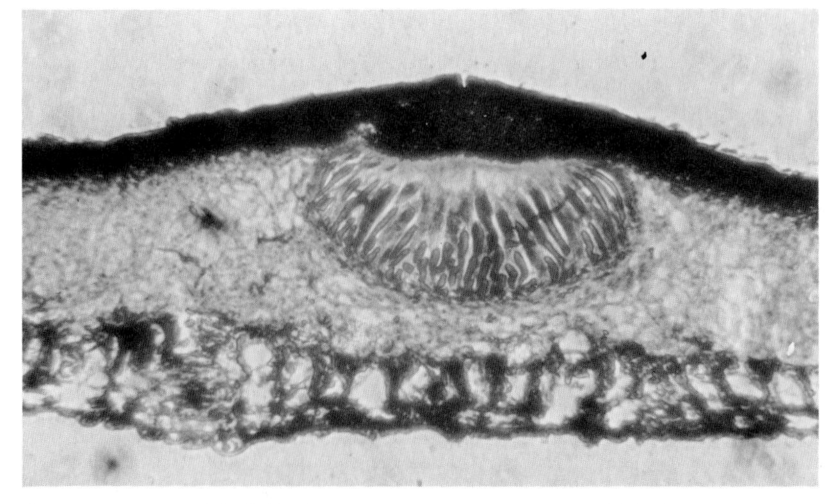

Fig. 9.25 Cross section of thick black stroma covering apothecium of *Rhytisma acerinum* within leaf tissues. (Photograph from microscope slide prepared by Carolina Biological Supply Company, Burlington, North Carolina.)

spots do not coalesce; instead they form discrete small black spots in groups (Fig. 9.23).

Etiology (R. acerinum): Germ tubes are presumed to enter the host through stomata, infecting epidermal and mesophyll cells (Fig. 9.24). As the fungus proliferates, infected cells rupture and a stroma with a thin black covering forms. Pycnidia develop within each stroma and release conidia, the function of which are not clearly understood. Further hyphal proliferation produces an apothecium with a thick black stroma over its surface (Fig. 9.25). Stromata then coalesce, forming large black spots. The apothecia complete their development during the winter and early spring within the fallen leaves. During the spring, the black stromata split open (Fig. 9.26) and elongate ascospores are forcibly discharged. Severe infections may cause premature leaf fall.

Control: The disease is not considered serious enough to warrant control efforts. The only practical means of control is to remove potential inoculum by raking and destroying infected leaves.

Selected References

Duravetz, J. S., and Morgan-Jones, J. F. (1971). Ascocarp development in *Rhytisma acerinum* and *R. punctatum*. *Can. J. Bot.* **49,** 1267–1272.

Jones, S. G. (1925). Life history and cytology of *Rhytisma acerinum* (Pers.) Fries. *Ann. Bot. (London)* **39,** 41–75.

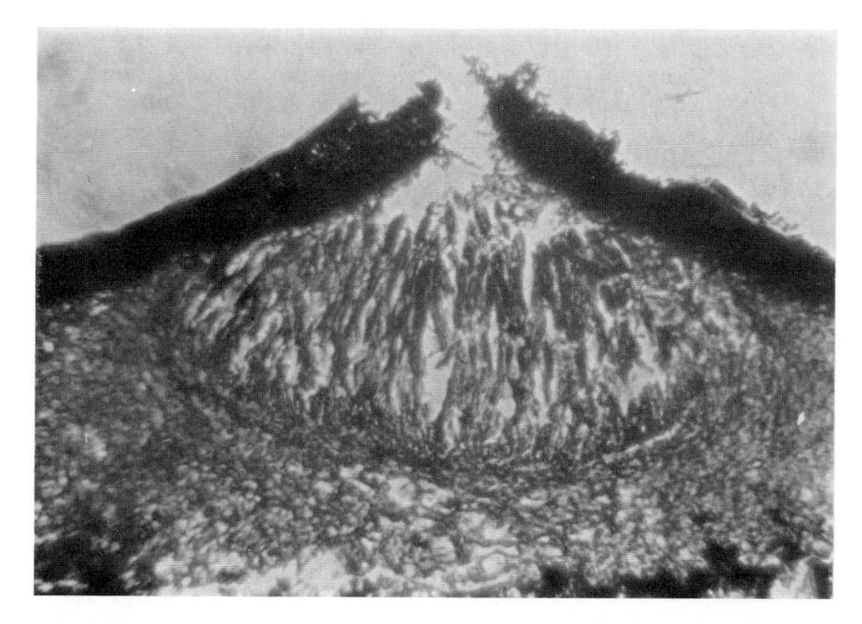

Fig. 9.26 Cross section of ruptured black stroma covering apothecium of *Rhytisma acerinum*. (Photograph from microscope slide prepared by Carolina Biological Supply Company, Burlington, North Carolina.)

DISEASE: NEEDLE CAST

Primary causal agents: Species of *Lophodermium, Scirrhia, Hypodermella, Bifusella, Adelopus, Rhabdocline, Elytroderma, Naemacyclus* and other related genera

Hosts: Pine (*Pinus* spp.), Fir (*Abies* spp.), Spruce (*Picea* spp.), and Douglas-fir (*Pseudotsuga menziesii* (Mirb.) Franco)

Symptomatology: Needle cast is a general term used to describe diseases in which needlelike leaves are shed from a host (Fig. 9.27). Specific needle casts can differ with respect to age of needles attacked and time of year affected, depending on the species of fungus and host infected. However, most needle casts have some common characteristics. Symptoms are first seen on the nee-

Fig. 9.27 Needle cast of spruce caused by *Lophodermium* spp. (Photograph courtesy of William E. MacHardy, University of New Hampshire, Durham.)

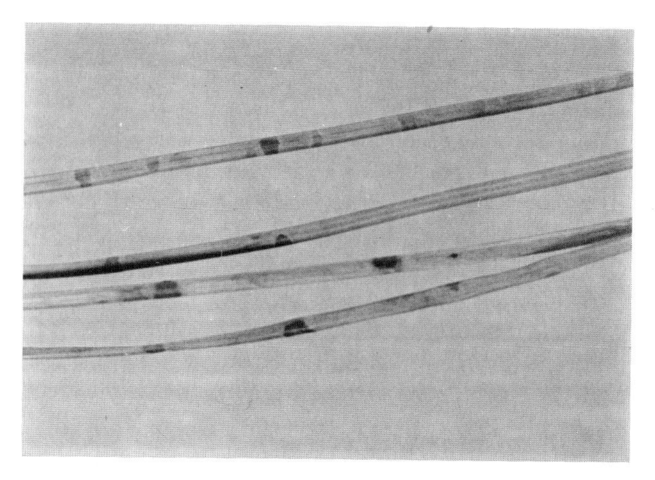

Fig. 9.28 Needle spot of pine caused by *Scirrhia* spp.

dles as light green to yellow spots that eventually turn red or brown (Fig. 9.28). Often the spots enlarge, encircling the needle and killing tissues beyond the spot. Movement of the pathogen into healthy parts of the leaf, or other infections on the same leaf cause death of the entire needle. Shedding then follows. Tiny, glossy black, elongated fruiting bodies can usually be found on one or both surfaces of infected needles (Fig. 9.29). These may be present

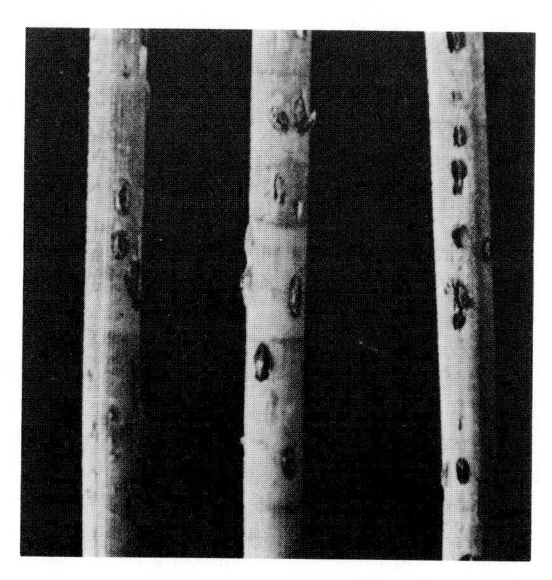

Fig. 9.29 Elongated apothecia on pine caused by *Lophodermium pinastri.* (Photograph courtesy of USDA Forest Service.)

while the needles are still on the tree or may develop after they have fallen to the ground.

Etiology (*Lophodermium pinastri* (Schrad. ex Fr.) Chev., taken as an example): Germ tubes of hyaline, threadlike ascospores enter needles through stomates during late summer or fall (Fig. 9.30). The fungus overwinters in the needles and resumes growth the following spring, causing brown spots with yellow margins. As fungal growth continues, the needles turn brown and are cast. Tiny black pycnidia with hyaline, elongate, and slightly curved conidia may be produced on the needles before they are cast. The function of conidia in development of the disease is not understood. Apothecia then develop on the dead needles, either on the tree or on the ground (Fig. 9.29). They appear as tiny, black, football-shaped structures with a slit down the middle. This type of apothecium has been more specifically termed a hysterothecium. The asci (Fig. 9.31) mature by late summer and ascospores are windblown to other susceptible hosts.

Control: Practical control measures for needle casts of forest trees are not known. However, trees in nurseries and in Christmas tree plantations can generally be treated by protectant fungicide applications during infection periods for effective control.

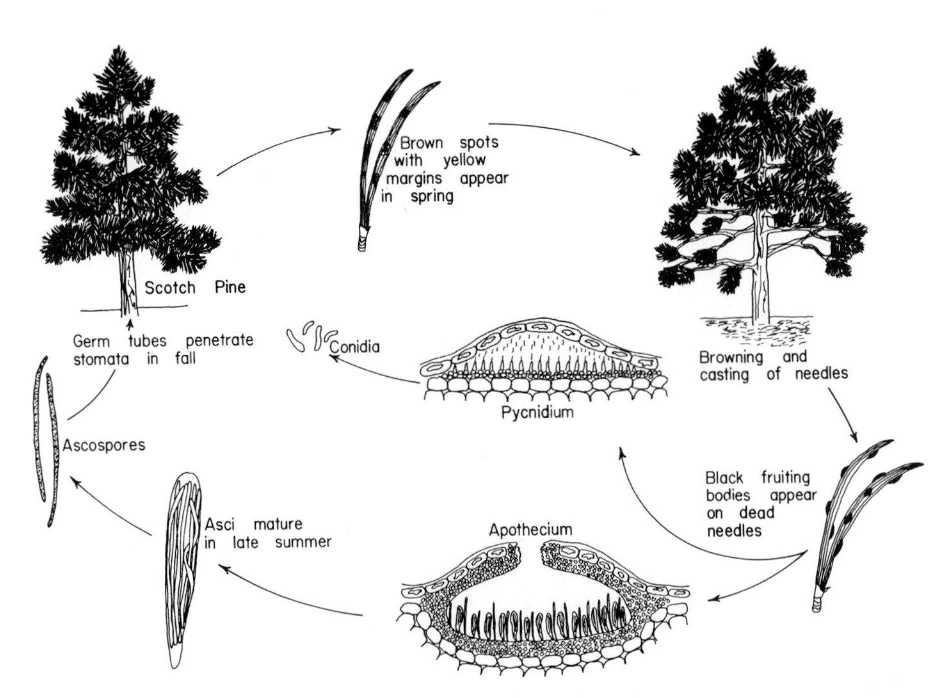

Fig. 9.30 Disease cycle of needle cast caused by *Lophodermium pinastri*.

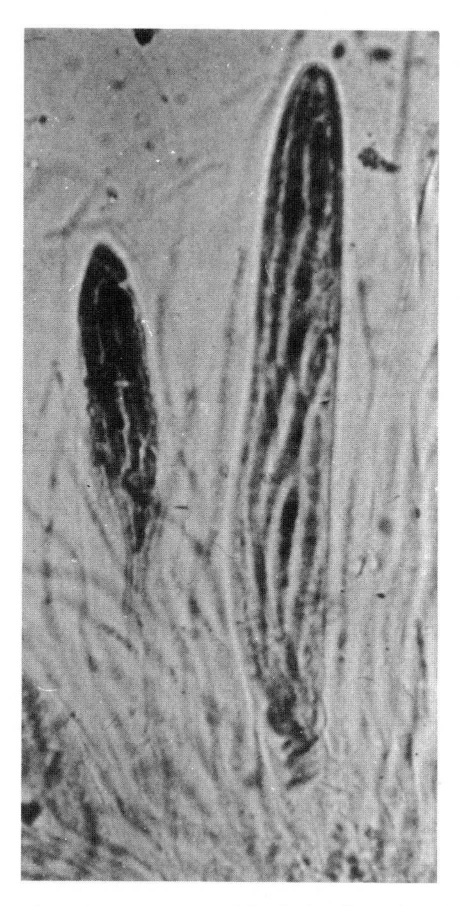

Fig. 9.31 Asci containing elongate ascospores of *Lophodermium pinastri.* (Photograph courtesy of USDA Forest Service.)

Selected References

Boyce, J. S., Jr. (1958). Needle cast of southern pines. *U.S. For. Serv., For. Pest Leafl.* No. 28.

Brandt, R. W. (1960). The *Rhabdocline* needle cast of Douglas-fir. *N.Y. State Univ. Coll. For. (Syracuse), Univ. Tech. Publ.* No. 84.

Childs, T. W. (1968). *Elytroderma* disease of ponderosa pine in the Pacific Northwest. *U.S. For. Serv., Res. Pap. PNW* **PNW-69.**

Darker, G. D. (1932). The Hypodermataceae of conifers. *Arnold Arboretum Contrib.* **1.**

Kistler, B. R., and Merrill, W. (1978). Etiology, symptomology, epidemiology, and control of *Naemacyclus* needlecast of Scotch pine. *Phytopathology* **68,** 267–271.

Merrill, W., and Longenecker, J. (1973). Swiss needle cast of Douglas-fir in Pennsylvania. *Plant Dis. Rep.* **57,** 984.

Nicholls, T. H., and Brown, H. D. (1972). How to identify *Lophodermium* and brown spot diseases on pine. *U.S. North Cent. For. Exp. Stn., Publ.* pp. 760–982.

Nicholls, T. H., and Skilling, D. D. (1974). Control of *Lophodermium* needle cast disease in nurseries and Christmas tree plantations. *U.S. For. Serv. Res. Pap. NC* **NC-110.**

DISEASE: NEEDLE AND TIP BLIGHT

Primary causal agents: Species of *Phomopsis, Diplodia, Rehmiellopsis, Ascochyta,* and other related genera

Hosts: Pine (*Pinus* spp.), Juniper (*Juniperus* spp.), Fir (*Abies* spp.), and Spruce (*Picea* spp.)

Symptomatology: Needle and tip blight, represents a broad category of diseases affecting coniferous species. Although needles may be cast during disease progression, the term blight usually signifies that other tissues in addition to leaf tissues are affected. Symptoms generally appear first on current year needles, which discolor and turn brown. Needles may droop, shrivel, or be cast, depending on the specific host and causal agent. Dieback of shoot tips follows, with lesions on larger stems frequently developing into cankers. Black fruiting structures of the pathogen often appear on dead needles and infected stems.

Etiology (*Phomopsis juniperovora* Hahn, taken as an example): Throughout the growing season, conidia are spread primarily by rainsplash to nearby susceptible host tissues where they germinate and penetrate nonwounded new foliage (Fig. 9.32). The fungus quickly colonizes the needles and moves into shoots and twigs. Small stems are usually girdled and killed. Larger stems develop cankers, with the fungus being confined to the cankered area. Pyc-

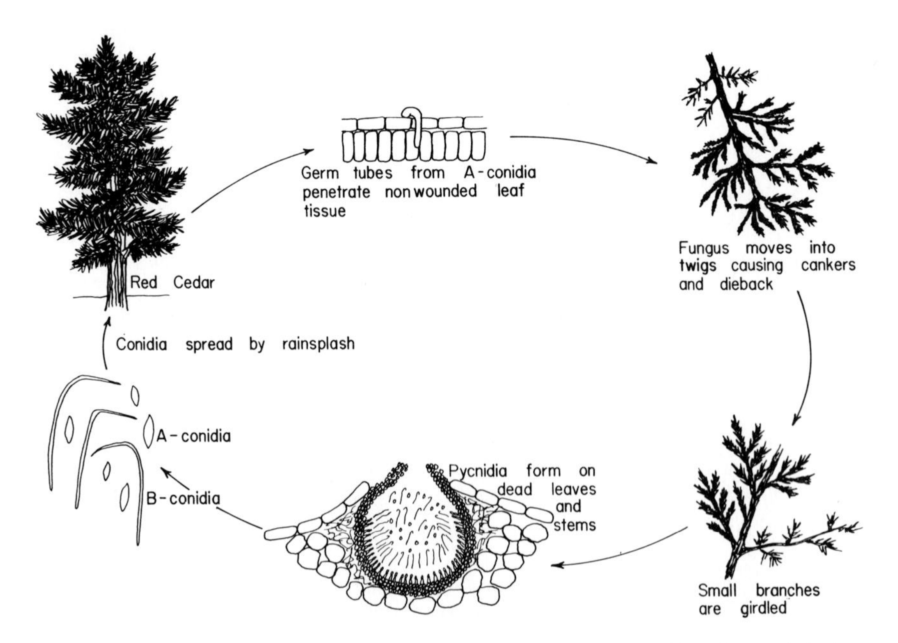

Fig. 9.32 Disease cycle of needle and tip blight of *Juniperus* spp. caused by *Phomopsis juniperovora*.

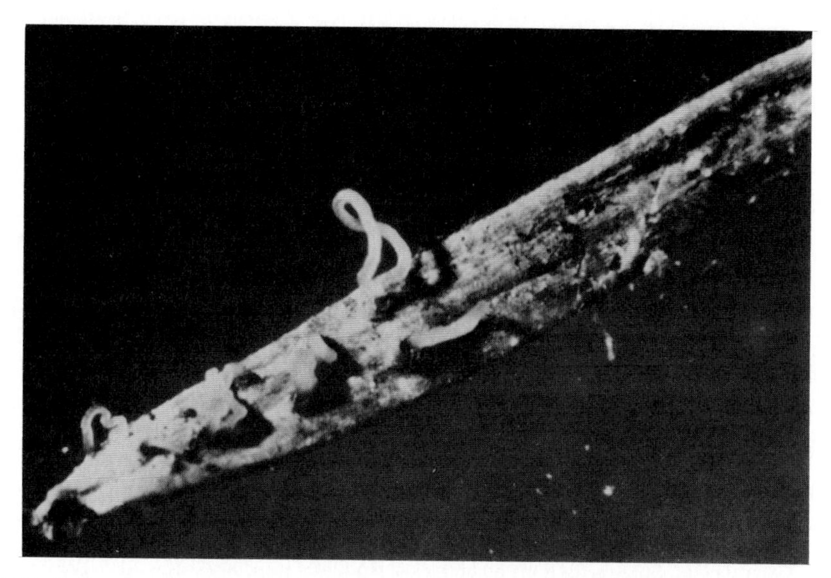

Fig. 9.33 Juniper needle with cirrhi oozing from pycnidia of *Phomopsis juniperovora*. (Photograph courtesy of Dr. Andrea Ostrofsky, University of New Hampshire, Durham.)

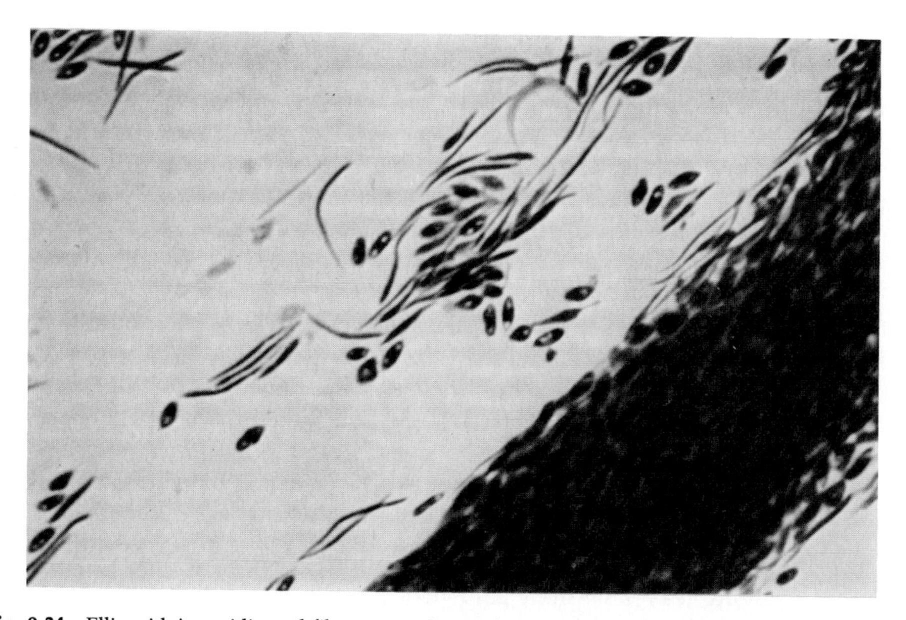

Fig. 9.34 Ellipsoid A-conidia and filamentous B-conidia to left of cirrhus of *Phomopsis juniperovora*. (Photograph courtesy of Dr. Andrea Ostrofsky, University of New Hampshire, Durham.)

nidia develop in the dead leaves and stems, releasing conidia in tendrils called cirrhi (Fig. 9.33). Two types of conidia are formed: A-conidia are hyaline, one-celled, and ellipsoid; B-conidia are hyaline, one-celled, filamentous, and curved (Fig. 9.34). Only the A-conidia germinate and infect new foliage. The fungus can remain viable in dead tissues and produce pycnidia for at least two years.

Control: Protectant chemicals are effective, but must be applied frequently and are only practical in nurseries or in small plantations. Pruning of infected branches will limit inoculum production, but should be done during dry weather. Since resistant varieties of several species susceptible to needle blight are available, these should be used whenever new plantings are made.

Selected References

Foster, J. A., Sinclair, W. A., and Johnson, W. T. (1973). *Diplodia* tip blight of pines. *Cornell Tree Pest Leafl.* **A-7.**

Gill, D. L. (1974). Control of *Phomopsis* blight of junipers. *Plant Dis. Rep.* **58,** 1012–1014.

Peterson, G. W. (1976). Some reflections on biological studies of needle diseases. *U.S. For. Serv., Res. Note RM* **RM-323.**

Peterson, G. W. (1977). Infection, epidemiology and control of *Diplodia* blight of Austrian, ponderosa, and Scots pines. *Phytopathology* **67,** 511–514.

Peterson, G. W., and Hodges, C. S., Jr. (1975). *Phomopsis* blight of junipers. *U.S. For. Serv., For. Pest Leafl.* No. 154.

Waterman, A. M. (1945). Tip blight of species of *Abies* caused by a new species of *Rehmiellopsis. J. Agric. Res.* **70,** 315–337.

DISEASE: SOOTY MOLD

Primary causal agents: Traditionally, ascomycetous fungi in the Family Capnodiaceae, but more accurately, many pigmented members of Ascomycetes and Deuteromycetes

Hosts: Most species of trees

Symptomatology: To call sooty mold a disease may be taking questionable liberties, but it is difficult to convince a homeowner that an *affected* tree is not *infected*. Leaf surfaces and branches appear as though coated with black soot, hence the name sooty mold (Fig. 9.35). This dark material consists of spores and mycelium of fungi. In most cases, the material may be easily washed or scraped off the affected parts. Evidence of sucking insects may be present (Fig. 9.36).

Etiology: Fungi that cause sooty mold are epiphytes, that is, they survive on the surface of living leaves without parasitizing them. Nourishment is obtained primarily from droplets of honeydew excreted by sucking insects. As the honeydew is deposited on the leaf, windblown spores (conidia and asco-

Fig. 9.35 Sooty mold on pine (A), spruce (B), and tulip poplar (C).

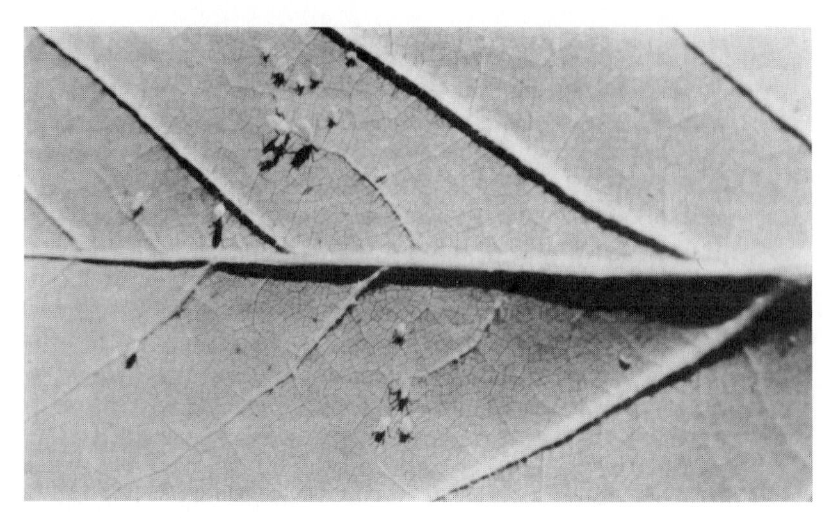

Fig. 9.36 Aphids feeding on lower leaf surface. (Photograph courtesy of William E. MacHardy, University of New Hampshire, Durham.)

spores) of one or more sooty molds germinate and are nourished by the complex mixture. Abundant dark mycelial growth develops, the amount dependent on the quantity of honeydew. In this sense, sooty mold is more an indicator of an insect problem than a disease problem. However, in severe cases, the dense material may reduce the amount of sunlight reaching the leaf surface.

Control: The most effective method of controlling sooty mold is to control sucking insects. Once this is accomplished, the black fungal material will eventually wash off branches and foliage by weathering.

Selected Reference

Hughes, S. J. (1976). Sooty moulds. *Mycologia* **68,** 693–820.

10

Wilt Diseases

INTRODUCTION

Wilt diseases are caused by fungi that rapidly disrupt the translocation stream resulting in severe moisture stress, wilting of the leaves, and often death of the tree in a short period. Of course, disease that blocks transpiration, such as a girdling stem canker or an extensive root rot, can also cause foliar wilting. But, it is the ability of the wilt disease fungi to invade the vascular tissues of living trees that makes this group unique. This invasion is usually recognized by the presence of discolored streaks or zones in the outer xylem. If the invasion of the wilt fungi is slow, foliar wilt may not occur, resulting in dieback and decline over long periods.

Wilt fungi gain entry to the susceptible tree in three major ways: (1) via feeding wounds of insect vectors on the branches, (2) through various wounds in the roots or buttress area, and (3) through living root grafts between adjacent trees of the same species. Dutch elm disease and oak wilt are two important examples of wilt diseases in which the fungi are transmitted by insect vectors, while Verticillium wilt and mimosa wilt are two important examples of wilt diseases in which soil-borne fungi enter through wounds in the roots or buttress area. All wilt pathogens can enter through root grafts. Location and method of entry by the wilt fungi determine the symptom patterns that result and the type of controls that can be used. Typically, fungi which cause diseases of the wilt type are either Ascomycetes or Deuteromycetes.

Fig. 10.1 Wilt in upper crown of American elm infected with *Ceratocystis ulmi.* (Photograph courtesy of Shade Tree Laboratories, University of Massachusetts, Amherst.)

Fig. 10.2 American elms showing severe symptoms of Dutch elm disease.

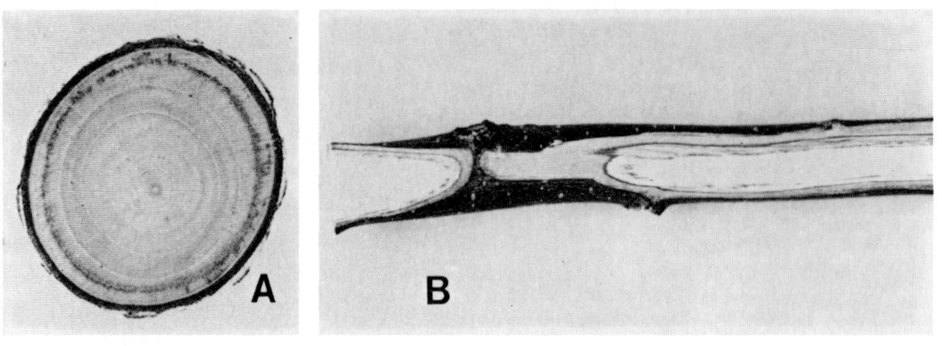

Fig. 10.3 American elm twigs infected with *Ceratocystis ulmi*. Discolored xylem in cross section (A) and longitudinal section (B).

DISEASE: DUTCH ELM DISEASE

Primary causal agent: *Ceratocytis ulmi* (Buism.) C. Mor.

Vectors: Native elm bark beetle (*Hylurgopinus rufipes* Eichh.), European elm bark beetle (*Scolytis multistriatus* (Marsh.))

Fig. 10.4 Egg and larval galleries of European elm bark beetle on American elm logs. (Photograph courtesy of Shade Tree Laboratories, University of Massachusetts, Amherst.)

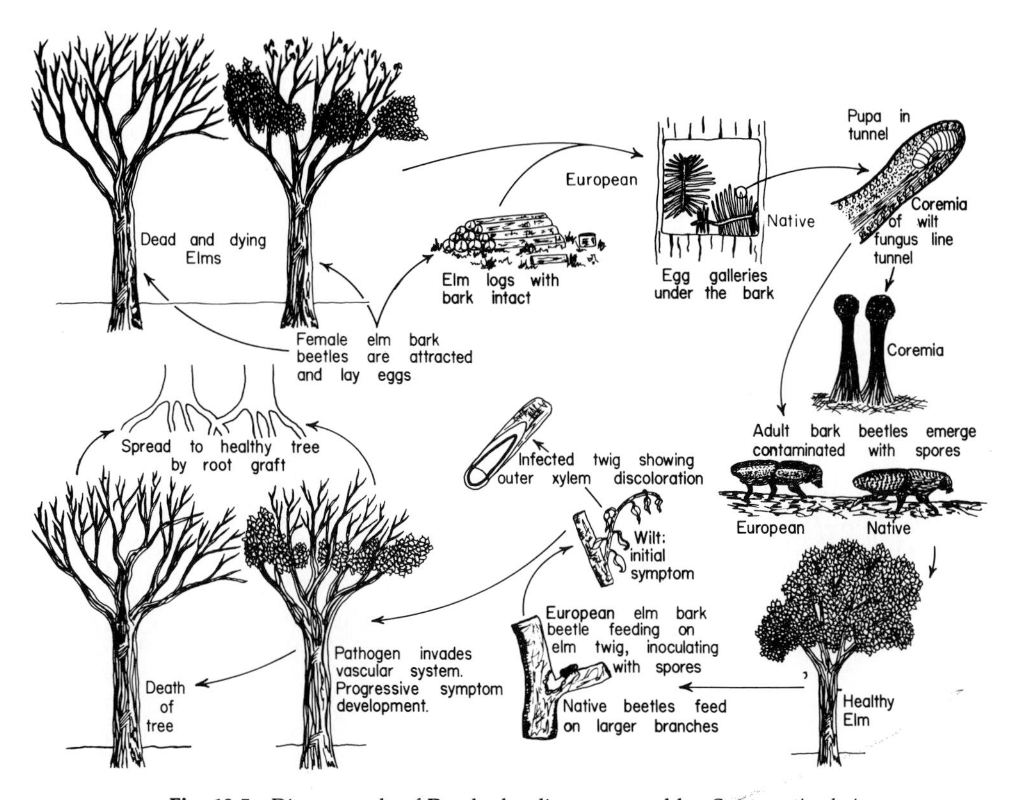

Fig. 10.5 Disease cycle of Dutch elm disease caused by *Ceratocystis ulmi*.

Hosts: Elms (*Ulmus* spp.) native to North America, such as American, red or slippery, September, and winged, are highly susceptible. European elms are moderately resistant. Asiatic elms are highly resistant.

Symptomatology: Yellowing and wilting occur on leaves of one to several branches in the upper crown during late spring or early summer (Fig. 10.1). The affected leaves quickly turn brown and die. These symptoms are repeated on progressively larger branches and often involve the entire tree by the end of the summer (Fig. 10.2). Infections later in the summer progress much slower, and only a few branches may be involved in the current season. However, further progression usually occurs in the next season. Severely infected trees often die in the current season and those surviving the winter usually are dead by the end of the next season. Affected branches contain brown discoloration in the outer xylem (Fig. 10.3). Discoloration can be found in twigs, large branches, the trunk, and sometimes in roots. In dead and dying trees egg and larval galleries of the insect vectors can often be found under the bark of the trunk and large branches (Fig. 10.4).

Etiology: The pathogen is carried to susceptible elms by insect vectors, elm bark beetles (Fig. 10.5). The insects transmit the pathogen during their feeding of either twigs [European elm bark beetle (Fig. 10.6A)] or larger branches [native elm bark beetle (Fig. 10.6B)]. The fungus enters the vascular system through the feeding wounds and proliferates in the xylem vessels. The invaded vessels become nonfunctional due to the activities of the pathogen, and xylem sap is no longer transported through them. The pathogen progressively invades the xylem vessels. Then, twigs, branches, and eventually whole sections of the crown are cut off from the transpiration stream, resulting in wilting, quickly followed by yellowing and browning of the leaves. Eventually the entire tree dies as the pathogen completely invades and kills the vascular tissue. Female elm bark beetles enter the bark of dead and dying elms and deposit eggs in galleries between the inner bark and xylem. Eggs hatch and the emerging larvae (Fig. 10.6C) tunnel under the bark and eventually pupate. When the adults emerge they are contaminated with spores of the *Graphium* conidial state (Fig. 10.7) which are produced in the beetle galleries. The adult beetles

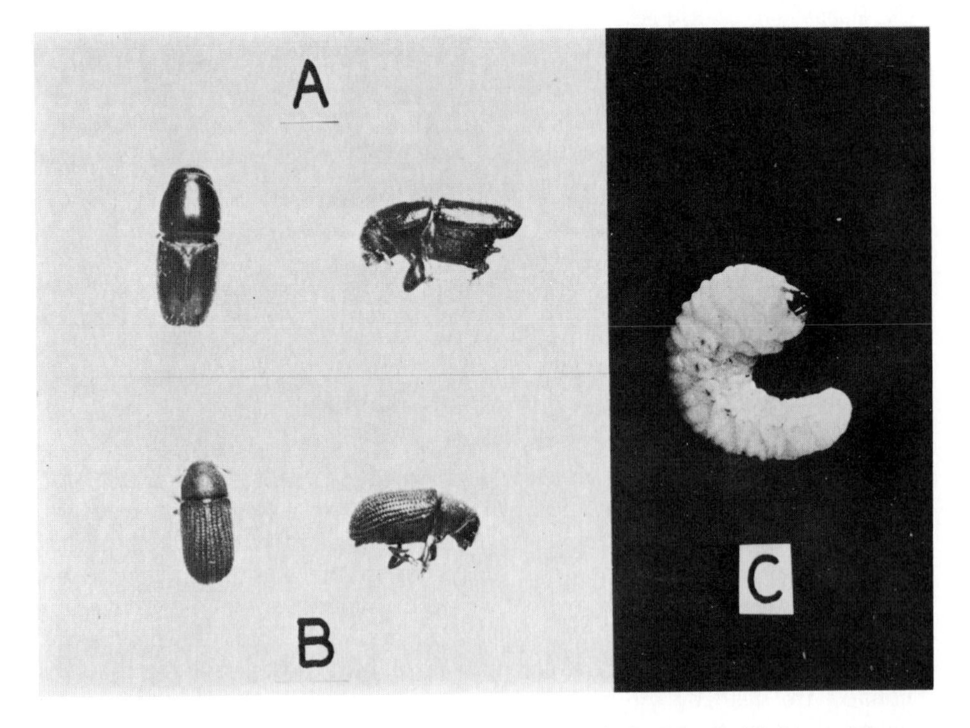

Fig. 10.6 Adult European elm bark beetle (A), adult native elm bark beetle (B), larva of European elm bark beetle (C). (Photographs courtesy of Shade Tree Laboratories, University of Massachusetts, Amherst.)

Fig. 10.7 Coremia of the *Graphium* state of *Ceratocystis ulmi*. (Photograph courtesy Botany and Plant Pathology Dept., University of New Hampshire, Durham.)

can then transmit the pathogen to healthy trees. The sexual state of the fungus, a long-necked perithecium, is rare in nature. However, it may play a role in the development of new strains of the pathogen. In laboratory culture (Fig. 10.8) a second conidial state, *Cephalosporium* spp., can be easily isolated from infected elm tissue (see Exercise IV). When elm trees of the same species grow close to each other, root grafts often form and the pathogen can move through the root system of an infected tree to invade an adjacent healthy one.

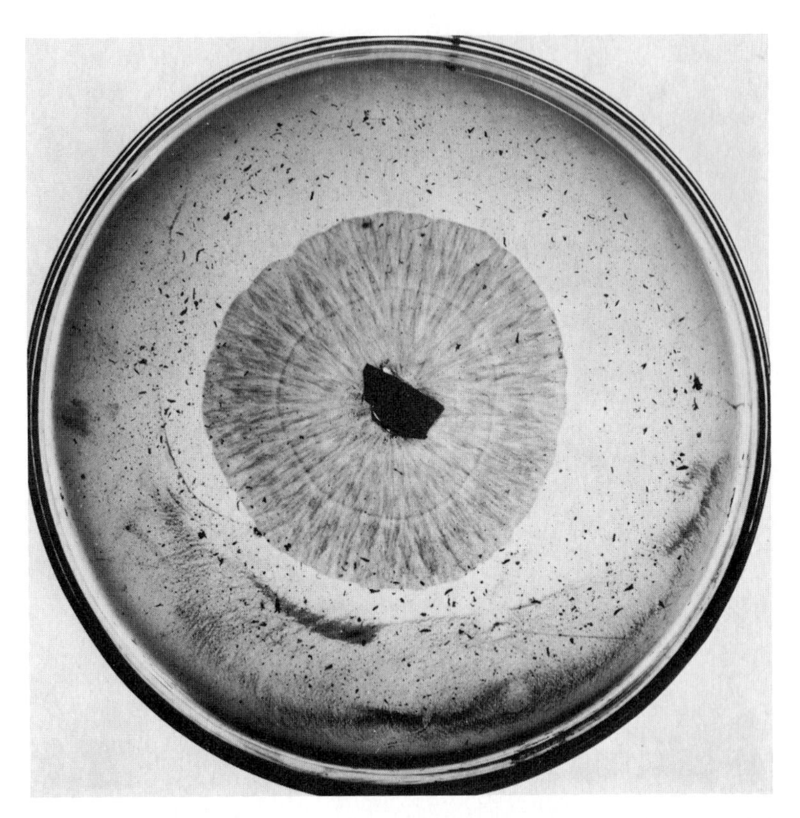

Fig. 10.8 Young culture of *Ceratocystis ulmi* grown from a section of discolored wood placed on a nutrient medium. (Photograph courtesy of Shade Tree Laboratories, University of Massachusetts, Amherst.)

Control:

Sanitation: Remove all dead and dying elms as soon as they are detected. Remove all infected branches during the initial stage of the disease by pruning at least 10 feet (3 meters) behind any vascular discoloration. Cut root grafts adjacent to infected trees either by trenching to 3 feet (1 meter) or with a soil fumigant to kill a thin band of roots.

Vector control: Apply dormant insecticide spray during early spring. Pheromones have been synthesized and show promise for reducing beetle populations by capturing them in pheromone traps.

Prophylaxis and therapy: Numerous methods for tree injection of materials inhibitory to the pathogen, to protect healthy trees or to save infected ones, are currently under investigation. The effectiveness of injection as a treatment for Dutch elm disease is controversial and there is concern over repeated multiple wounding from injections.

Selected References

Banfield, W. M. (1968). Dutch elm disease recurrence and recovery in American elm. *Phytopathol. Z.* **62,** 21–60.

Neely, D., and Himelick, E. B. (1965). Effectiveness of Vapam in preventing root graft transmission of Dutch elm disease fungus. *Plant Dis. Rep.* **49,** 106–108.

Schreiber, L. R., and Townsend, A. M. (1976). Variability in aggressiveness, recovery, and cultural characteristics of isolates of *Ceratocystis ulmi. Phytopathology* **66,** 239–244.

Sinclair, W. A., and Campana, R. J., eds. (1978). Dutch elm disease—Perspectives after 60 years. *Cornell Agric. Exp. Stn., Publ.* **8,** No. 5.

Sinclair, W. A., Zahand, J. P., and Melching, J. B. (1975). Localization of injection in American elms resistant to *Ceratocystis ulmi. Phytopathology* **65,** 129–133.

Stipes, R. J., and Campana, R. J., eds. (1981). "A Compendium of Elm Diseases." Am. Phytopathol. Soc., St. Paul, Minnesota.

DISEASE: OAK WILT

Primary causal agent: *Ceratocystis fagacearum* (Bretz) Hunt

Vectors: Sap feeding beetles (Nitidulidae), Bark beetles (Scolytidae)

Hosts: All oaks (*Quercus* spp.) are susceptible. Species in the red oak group are highly susceptible. Species in the white oak group are moderately resistant.

Symptomatology:

Red Oak Group: In mid to late spring leaves turn dull green initially and then yellow or brown from the tip progressing toward the petiole. Leaf abscission occurs simultaneously and leaves are shed in all stages of symptom development, usually beginning at the top and moving downward (Fig. 10.9). After most of the leaves have been dropped, sucker growth often appears on the trunk and major branches. Death of the tree often occurs by the end of the summer but can occur sooner. Discoloration of outer xylem of twigs and branches is rarely found in the red oak group. In dead trees bark cracks may appear which overlie raised pads of fungus mycelium on the outer xylem.

White Oak Group: Similar foliar symptoms occur but are restricted to a small number of branches each year. Progressive decline occurs over several seasons and the tree eventually dies, although some trees in the white oak group may recover. Discoloration in the outer xylem of twigs and branches is sometimes found (Fig. 10.10). In dead trees, bark cracks and fungus pads may occur as in the red oak group.

Etiology: The pathogen is transmitted to healthy susceptible oaks by insect vectors or through root grafts with adjacent infected trees (Fig. 10.11). Fungus-contaminated insect vectors introduce *C. fagacearum* into the xylem through fresh mechanical wounds or through feeding wounds. Invasion by the pathogen through the xylem causes vessel dysfunction and subsequent

Fig. 10.9 Defoliation and sucker growth of red oak infected with *Ceratocystis fagacearum*. (Photograph courtesy of Botany and Plant Pathology Dept., University of New Hampshire, Durham.)

Fig. 10.10 Discoloration in outer xylem of white oak twig infected with *Ceratocystis fagacearum*. (Photograph courtesy of Botany and Plant Pathology Dept., University of New Hampshire, Durham.)

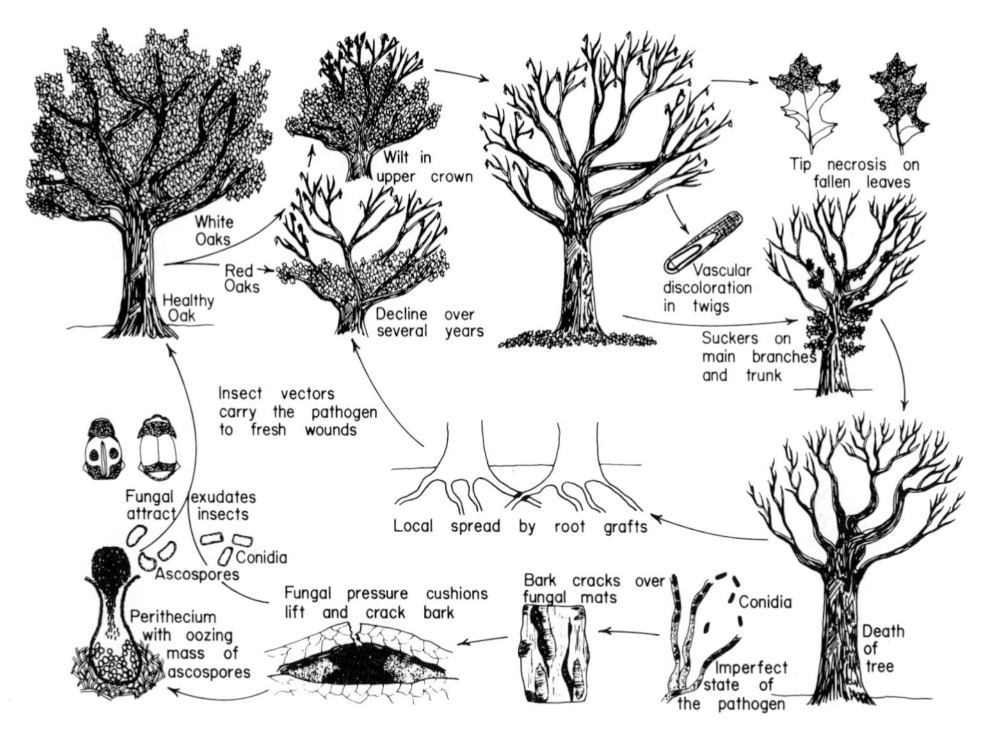

Fig. 10.11 Disease cycle of oak wilt caused by *Ceratocystis fagacearum*.

blockage of the transpiration stream. Continued vascular invasion results in the death of the tree. The pathogen is able to extensively invade the xylem after death and produce fungal mats, which exert outward pressure under the bark resulting in cracking. Insect vectors are attracted to the sweet odor of the fungal mats and enter through the bark cracks. Their bodies become contaminated with conidia of the pathogen. Oak wilt has a similar etiology in both the red and white oak groups.

Control: Avoid wounds, even pruning wounds, that would attract insect vectors during early spring to midsummer. Cover any wounds made during this period with wound dressing to protect wounds from *C. fagacearum* contamination. Sever root grafts between healthy and infected trees mechanically or with a soil fumigant. Prune affected branches of white oaks, and remove infected red oaks as soon as possible.

Selected References

Jones, T. W. (1971). An appraisal of oak wilt control programs in Pennsylvania and West Virginia. *USDA For. Serv., Res. Pap. NE* **NE-204.**

Jones, T. W., and Phelps, W. R. (1972). Oak wilt. *U.S. For. Serv., For. Pest Leafl.* No. 29.

Rexrode, C. O., and Jones, T. W. (1970). Oak bark beetles—Important vectors of oak wilt. *J. For.* **68,** 294–297.
True, R. P., Barnett, H. L., Dorsey, C. K., and Leach, J. G. (1960). Oak wilt in West Virginia. *W. Va. Agric. Exp. Stn. Bull.* No. 448T.

DISEASE: VERTICILLIUM WILT

Primary causal agent: *Verticillium dahliae* Kleb.

Hosts: Most deciduous trees are susceptible. Some resistant species include beech, birch, oak, sycamore, and willow. Conifers are immune.

Symptomatology: Affected trees often exhibit both wilt and dieback (Fig. 10.12). Acute wilt symptoms result in a rapid collapse of foliage on a branch or a

Fig. 10.12 Wilt of sugar maple (A), wilt and dieback of Japanese maple (B), closeup of B (C) caused by *Verticillium dahliae.* (Photograph A courtesy of Shade Tree Laboratories. Photographs B and C courtesy of George N. Agrios, University of Massachusetts, Amherst.)

section of the crown. These symptoms are often preceded by leaf stunting, infolding of leaf margins (cupping) and occasional leaf scorch, and heavy seed production. Dieback follows the wilt symptoms on an affected branch but also occurs when affected branches do not leaf in spring. Elongate cankers can also form on the main stem around affected branches or separately on the trunk. Bark cracking often occurs on the canker surface and a brown-black ooze may flow from the canker margins. Sucker growth is common below the cankers and below any dead sections of the crown. In addition, branches below these

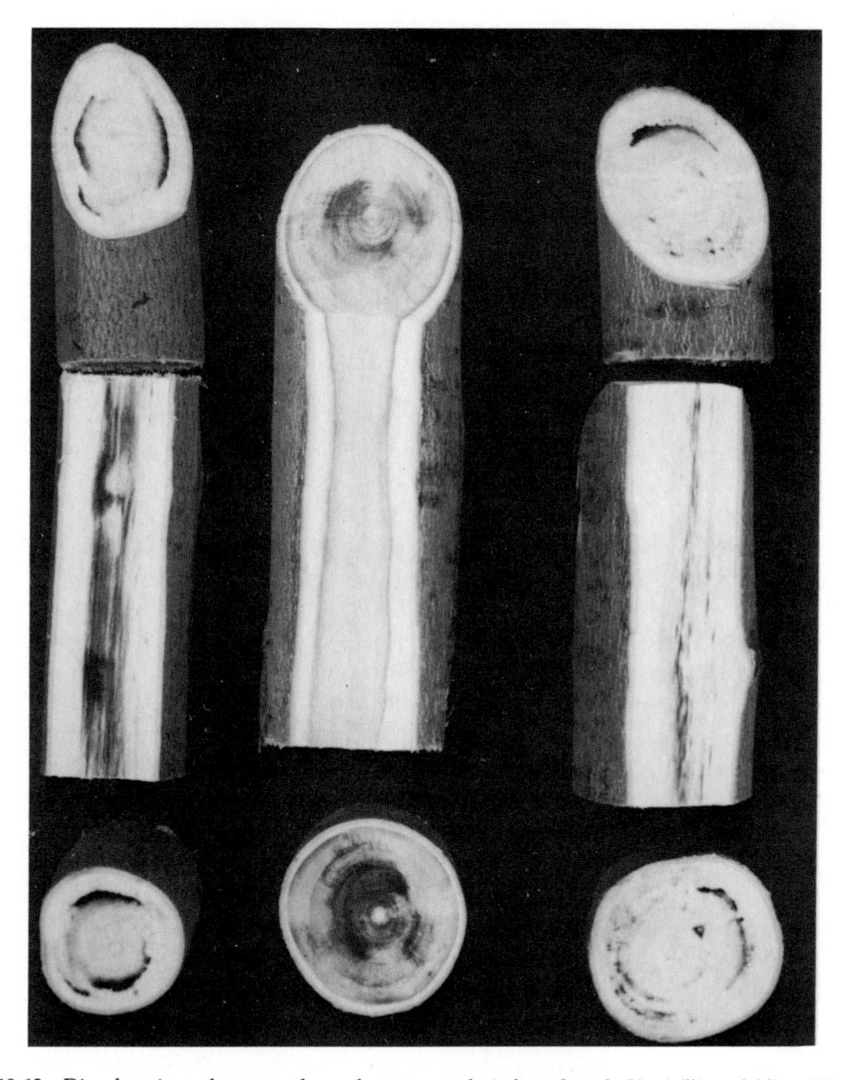

Fig. 10.13 Discoloration of outer xylem of sugar maple infected with *Verticillium dahliae.* (Photograph courtesy of Wayne A. Sinclair, Cornell University, Ithaca, New York.)

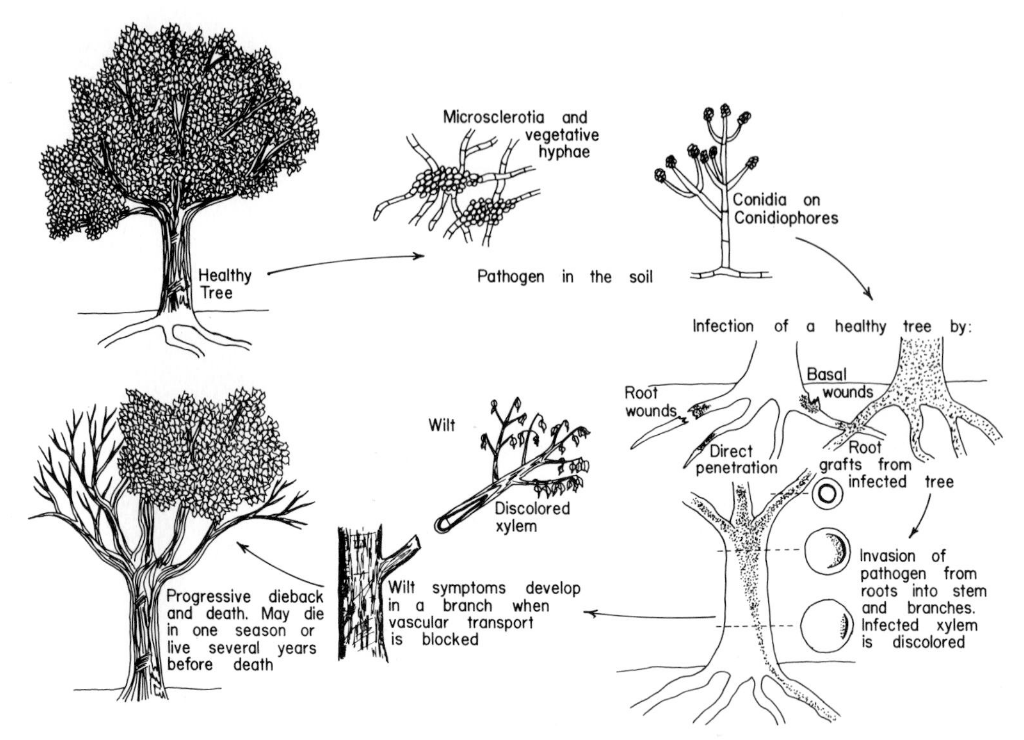

Fig. 10.14 Disease cycle of Verticillium wilt caused by *Verticillium dahliae.*

regions often exhibit abnormally large foliage. The outer xylem of affected branches and the trunk exhibit vascular discoloration (Fig. 10.13), the color of which is host specific. For example, maples exhibit green-black discoloration, while black locust will be brown to black, and elm will be brown. Xylem discoloration associated with Verticillium wilt is most common in the trunk and larger branches and may not reach small branches, especially in larger trees. Discoloration can be followed into the roots of infected trees, but the color is diminished and it is confined to smaller areas.

Etiology: The pathogen enters wounds on the roots or buttress, and moves through the xylem into the trunk (Fig. 10.14). It invades the outer xylem up the tree, remaining mostly in the current year's vessels and extending radially as it moves vertically. Pathogen activities in the xylem cause dysfunction and discoloration of the vascular tissue. Wilt occurs when most of the vessels supplying a branch or portion of the crown are no longer functional (Fig. 10.15). Continued vascular invasion of *V. dahliae* will result in progressive wilt and branch dieback until death occurs. If the tree survives the summer, the pathogen will often reinvade through the roots the following season. Death

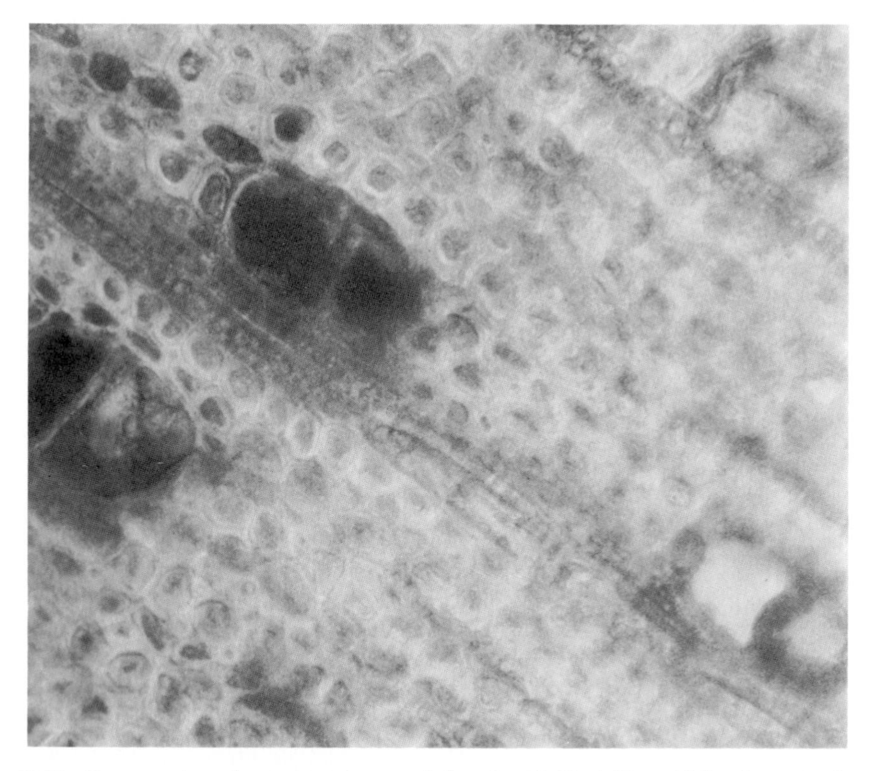

Fig. 10.15 Cross section of sugar maple stem infected with *Verticillium dahliae* showing plugged vessels (upper left) and open vessels (lower right). 800×.

often occurs within 2–3 years. Occasionally, trees, usually larger ones, will survive the disease for many years with minor symptoms each year or even without further symptom development. The pathogen persists in the soil as vegetative mycelium on plant debris or as microsclerotia (Fig. 10.16), a long-lived vegetative resting stage.

Control: Avoid replanting a susceptible species when a tree has been killed by Verticillium wilt, or fumigate the soil before replanting. In cases with minor symptom development improve vigor by fertilizing and watering, and remove any dieback to improve appearance.

Selected References

Born, G. L. (1974). Root infection of woody hosts with *Verticillium albo-atrum*. *Bull.—Ill. Nat. Hist. Surv.* **31,** 205–249.

Caroselli, N. E. (1957). Verticillium wilt of maples. *R.I. Agric. Exp. Stn., Bull.* No. 335.

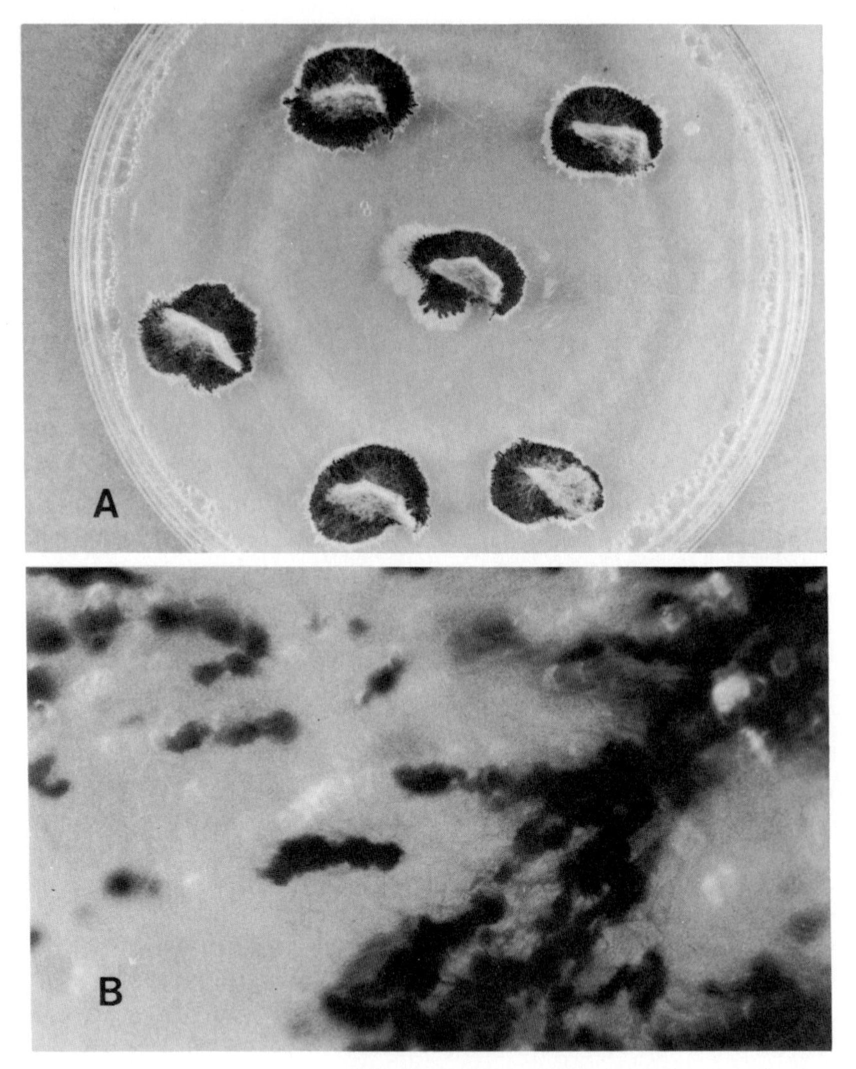

Fig. 10.16. Agar culture of *Verticillium dahliae* showing mycelium and dark microsclerotia around wood chips from an infected tree (A), closeup (B). 50×. (Photograph A courtesy of George N. Agrios, University of Massachusetts, Amherst.)

Himelick, E. B. (1969). Tree and shrub hosts of *Verticillium albo-atrum*. *Ill. Nat. Hist. Surv., Publ.* **66.**

Malia, M. E., and Tattar, T. A. (1978). Electrical resistance, physical characteristics, and cation concentrations in xylem of sugar maple infected with *Verticillium dahliae*. *Can. J. For. Res.* **8,** 322–327.

Sinclair, W. A., Smith, K. L., and Larsen, A. O. (1981). Verticillium wilt of maples: Symptoms related to movement of the pathogen in stems. *Phytopathology* **71,** 340–345.

Fig. 10.17 Wilted leaves of mimosa infected with *Fusarium oxysporum* f. *perniciosum* (Photograph courtesy of Robert Anderson, USDA Forest Service, Asheville, North Carolina.)

Fig. 10.18 Dieback in crown of mimosa infected with *Fusarium oxysporum* f. *perniciosum* (Photograph courtesy of Robert Anderson, USDA Forest Service, Asheville, North Carolina.)

DISEASE: MIMOSA WILT

Primary causal agent: *Fusarium oxysporum* (Schl.) em. Synd. & Hans. f. *perniciosum* (Hept.) Toole

Hosts: Mimosa or Silk tree (*Albizzia julibrissin* Duraz.)

Symptomatology: Leaves wilt (Fig. 10.17), turn yellow, die, and then abscise in rapid succession, one branch at a time until the entire crown is dead (Fig. 10.18). Bark splitting with subsequent ooze flux (Fig. 10.19) may occur during the advanced stages of the disease. Death usually occurs by the end of the current growing season. Internal discoloration of the outer xylem occurs in patterns similar to Verticillium wilt and is most easily detected in the roots and the buttress area (Fig. 10.20).

Fig. 10.19. Bark splitting and ooze flux on mimosa infected with *Fusarium oxysporum* f. *perniciosum* (A), closeup (B). (Photographs courtesy of Robert Anderson, USDA Forest Service, Asheville, North Carolina.)

Fig. 10.20 Discoloration in outer xylem in lower stem of mimosa infected with *Fusarium oxysporum* f. *perniciosum*. (Photograph courtesy of Robert Anderson, USDA Forest Service, Asheville, North Carolina.)

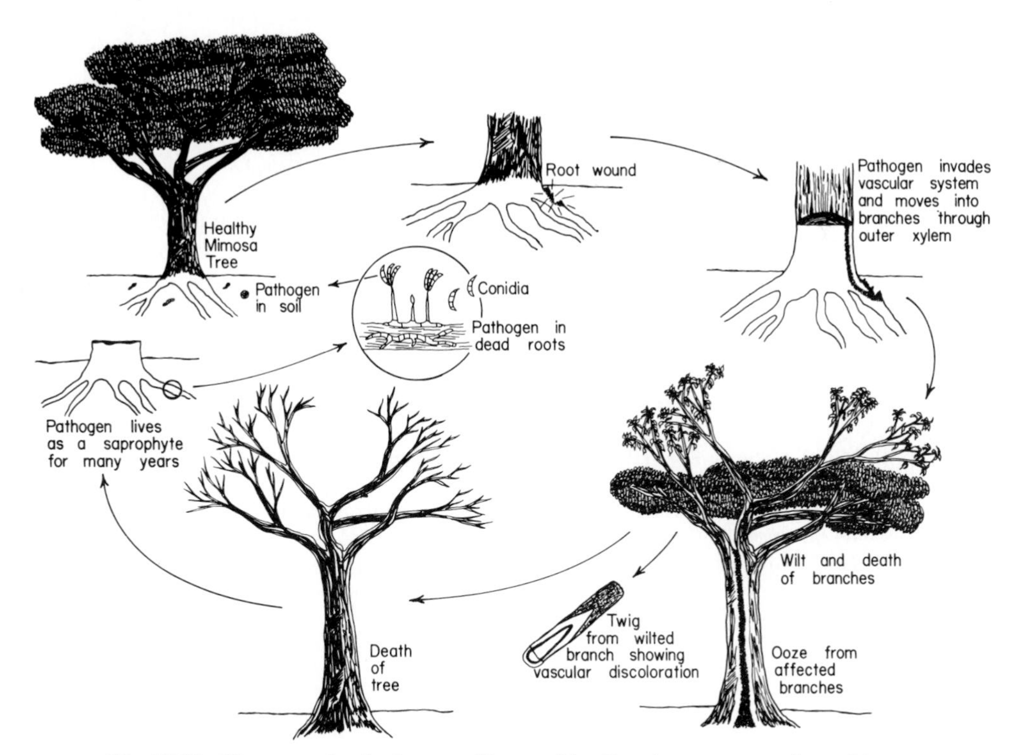

Fig. 10.21 Disease cycle of mimosa wilt caused by *Fusarium oxysporum* f. *perniciosum*.

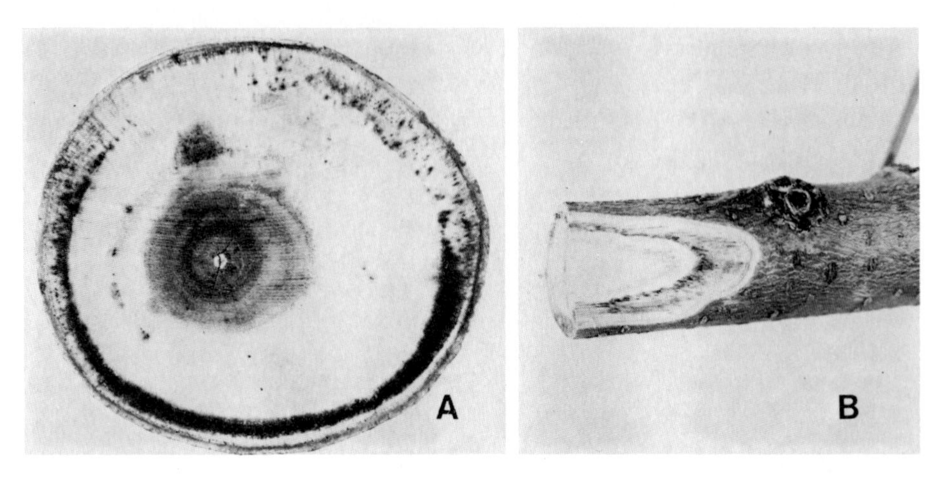

Fig. 10.22 Xylem discoloration in twigs of mimosa, cross section (A), longitudinal section (B) caused by *Fusarium oxysporum* f. *perniciosum*.

Etiology: The pathogen is a soil-borne fungus that enters the susceptible tree through wounds in the roots or buttress area (Fig. 10.21). The fungus moves through the outer xylem vessels and causes vascular dysfunction and induces xylem discoloration (Fig. 10.22). Wilt symptoms occur in each branch when most of the vascular tissues leading to it are no longer functional. Progressive vascular invasion results in death of the entire tree. The pathogen may persist as a saprophyte in dead host tissue or in the soil indefinitely.

Control: Use resistant varieties of mimosa. Avoid replanting the species where a mimosa has died from mimosa wilt unless the dead root system has been removed and the soil has been fumigated prior to planting.

Selected References

Hepting, G. H. (1939). A vascular wilt of the mimosa tree (*Albizzia julibrissin*). *U.S. Dep. Agric., Circ.* No. 535.

Phipps, P. M., and Stipes, R. J. (1975). Control of Fusarium wilt of mimosa with benomyl and thiabendazole. *Phytopathology* **65,** 504–506.

Stipes, R. J., and Phipps, P. M. (1975). *Fusarium oxysporum* f. *perniciosum* on *Fusarium*-wilted mimosa trees. *Phytopathology* **65,** 188–190.

Toole, E. R. (1941). *Fusarium* wilt of the mimosa tree (*Albizzia julibrissin*). *Phytopathology* **31,** 599–616.

11

Canker Diseases

INTRODUCTION

The term "canker" refers to a symptom caused by the death of definite and generally localized areas of bark and cortex on branches or trunks of trees. Most trees are susceptible to one or more canker-causing agents. These may be classified as noninfectious (abiotic) agents such as frost and sunscald, or infectious (biotic) agents such as bacteria, viruses, and fungi. Most cankers are caused by fungi in the class Ascomycetes.

Fungi that cause cankers are primarily passive opportunists, that is, they do not actively invade host tissues, but rather, they wait for an "opportunity" to invade through wounds or areas of stress. The bark of trees serves as a natural barrier to invasion, so therefore most canker-causing fungi must enter the host through wounds and to a lesser extent, natural openings, such as stomata, lenticels, and leaf scars. Little is known about the exact mechanism of how the invading fungi kill and colonize host tissues, but evidence suggests that some fungi produce toxins that kill the tissues ahead of the advancing mycelium. Cankers are usually grouped into one of the following three types:

1. Annual Canker. For this type of canker the fungus becomes established during the dormant season of the host, but during the following growing season, the host produces callus tissues sufficient to prevent further spread. Such

cankers are of little concern because damage is usually minimal and may be completely overlooked. In fact, in products such as lumber and veneer, an annual canker may only appear as a small dark streak in the wood.

2. Perennial Canker. Cankers of this type are more important from a pathological viewpoint than annual cankers, because an invading fungus may survive indefinitely in the host. Perennial cankers are seldom responsible for host mortality, primarily because there is a balanced interaction between the host and the pathogen. During the dormant season of the host, the mycelium of the pathogen becomes active and invades and kills the newly developed tissues. During the next growing period the host lays down new callus tissues ahead of the advancing mycelium. The fungus survives saprophytically in the dead tissues until late in the season, when it again advances through the newly formed dormant tissues. The previously formed callus tissues are usually killed as well as the adjacent tissues. Alternate formation and killing of callus tissues often gives such cankers a "targetlike" appearance. In some cases, the fungus is unable to kill callus in certain parts of the canker. This usually results in a canker that has an irregular shape. Occasionally the fungus causing a perennial canker may be held in check after several years of growth, and the infected area may be completely callused over. Although perennial canker-causing fungi may not actually be responsible for killing a tree, the cankered area itself is usually weakened, making it susceptible to wind breakage.

3. Diffuse Canker. This is the most serious type of canker for a tree, primarily because host mortality is usually the final outcome. For diffuse canker formation, the invading fungus usually grows vigorously during the growing season along with the host. This type of canker formation results in the host having little chance to produce an arresting callus, and girdling results within a few years.

DISEASE: CHESTNUT BLIGHT

Primary causal agent: *Endothia parasitica* (Murr.) A. & A.

Hosts: Primarily American chestnut (*Castanea dentata* (Marsh.) Borkh.), European chestnut (*C. sativa* Mill.), and chinkapins (*Castanea* spp.). Occasionally post oak (*Quercus stellata* Wang) and live oak (*Q. virginiana* Lam.).

Symptomatology: The first noticeable symptom is usually a "flag," a dead branch with yellow or brown wilted leaves. A diffuse girdling canker can usually be found on the branch below the discolored foliage (Fig. 11.1). Water sprouts frequently develop just below the canker. Young cankers on smooth-barked stems are yellowish-brown in color and may become sunken or swollen as the canker develops (Fig. 11.2). Mycelial fans can be seen beneath the cankered bark (Fig. 11.3). Orange pycnidia cover the surface of the canker and, during wet weather, exude spores in long twisting cirrhi (Fig. 11.4). Clusters of black perithecia form within the same tissues as the pycnidia.

Fig. 11.1 Diffuse girdling canker on American chestnut caused by *Endothia parasitica*. Note water sprouts below canker. (Photograph courtesy of Shade Tree Laboratories, University of Massachusetts, Amherst.)

Etiology: The fungus enters the host through wounds in the bark (Fig. 11.5). Germ tubes that enter these wounds are produced by both conidia and ascospores. The fungus spreads through the bark cortex, forming mycelial fans that penetrate the inner bark and kill the cambium. The outer bark over the diseased tissue swells and eventually begins to split. Orange pycnidia are formed

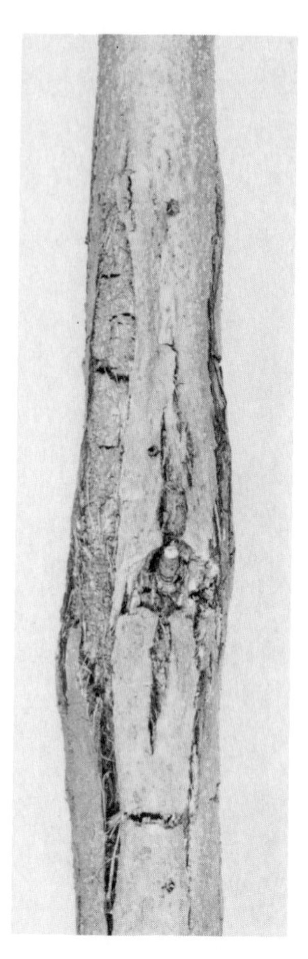

Fig. 11.2 Swollen canker on American chestnut caused by *Endothia parasitica*. Note abundant bark cracking around canker.

from the fungal tissue and push up through the epidermis of the bark. One-celled conidia are exuded from the pycnidia in long cirrhi, held together by a gelatinous matrix. The conidia are passively dispersed by rainsplash, or actively dispersed by insects, birds, and other animals. After pycnidia are produced, clusters of black perithecia form in the same fungal tissues. These perithecia are embedded in a stroma and have long necks that extend to the bark surface (Fig. 11.6). Bicellular ascospores are forcibly discharged and carried by the wind to new hosts. The fungus continues to grow in a susceptible host until the tree is girdled.

Control: Development of disease-resistant varieties is and has been the major control effort associated with American chestnuts. Unfortunately, those va-

Fig. 11.3 Mycelial fans of *Endothia parasitica* beneath bark of American chestnut. (Photograph courtesy of Shade Tree Laboratories, University of Massachusetts, Amherst.)

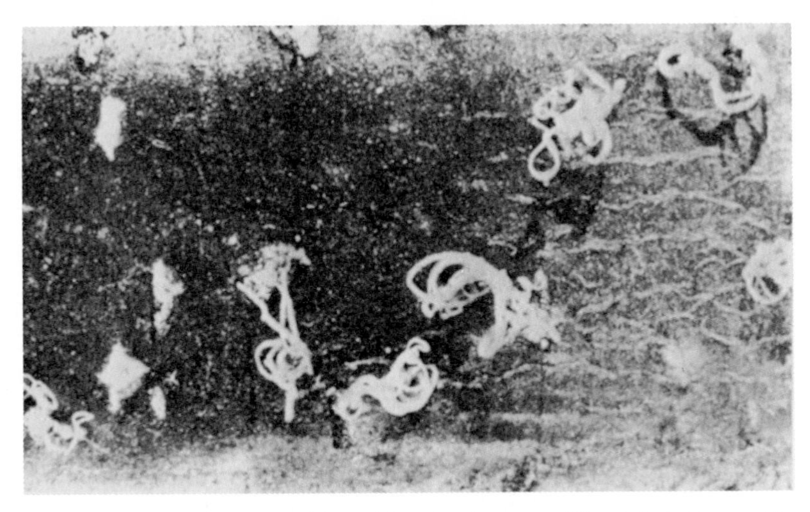

Fig. 11.4 American chestnut stem with cirrhi oozing from pycnidia of *Endothia parasitica*. (Photograph courtesy of USDA Forest Service.)

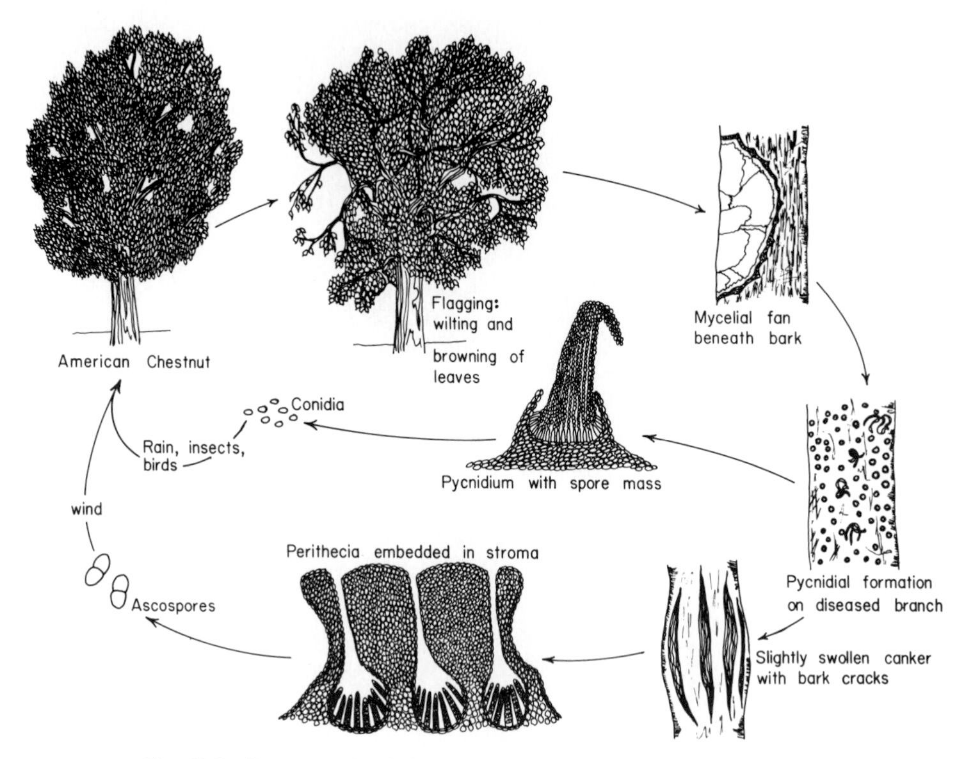

Fig. 11.5 Disease cycle of chestnut blight caused by *Endothia parasitica.*

rieties that have been resistant do not have the desired growth characteristics of the American chestnut. Natural development of a resistant tree is hindered by the fact that sprout trees are often killed before reaching sexual maturity, thus limiting genetic variation. Recent experiments with systemic chemicals and the introduction of hypovirulent strains of the fungus have shown some success as potential control measures for high value trees.

Selected References

Anagnostakis, S. L., and Jaynes, R. A. (1973). Chestnut blight control: Use of hypovirulent cultures. *Plant Dis. Rep.* **57,** 225–226.

Day, P. R., Dodds, J. A., Elliston, J. E., Jaynes, R. A., and Anagnostakis, S. L. (1977). Double-stranded RNA in *Endothia parasitica. Phytopathology* **67,** 1393–1396.

Diller, J. D. (1965). Chestnut blight. *U.S. For. Serv., For. Pest Leafl.* No. 94.

Hepting, G. H. (1974). Death of the American chestnut. *J. For. Hist.* **18,** 61–67.

Jaynes, R. A., and VanAlfen, N. K. (1974). Control of American chestnut blight by trunk injection with methyl-2-Benzimidazole carbamate (MBC). *Phytopathology* **64,** 1479–1480.

MacDonald, W. L., Cech, F. C., Luchok, J., and Smith, C., eds. (1979). *Proc. Am. Chestnut Symp., Morgantown, W. Va.*

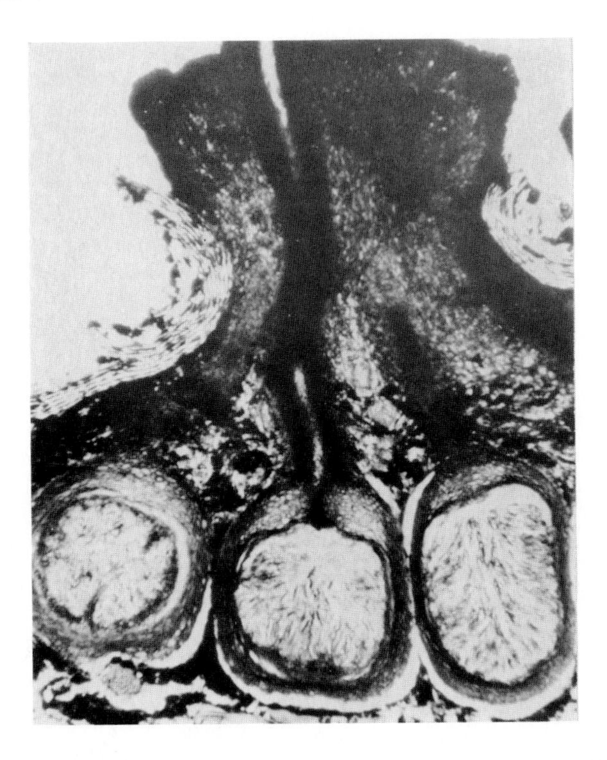

Fig. 11.6 Long-necked perithecia of *Endothia parasitica* embedded in stroma beneath the bark of American chestnut. (Photograph from microscope slides prepared by Triarch, Inc., Ripon, Wisconsin.)

DISEASE: STRUMELLA CANKER

Primary causal agent: *Strumella coryneoidea* Sacc. & Wint.

Hosts: Most commonly on red oak (*Quercus rubra* L.) and black oak (*Q. nigra* L.). Occasionally on basswood (*Tilia americana* L.), American beech (*Fagus grandifolia* Ehrh.), red maple (*Acer rubrum* L.), and shagbark hickory [*Carya ovata* (Mill.) K. Koch].

Symptomatology: Young cankers are difficult to detect on rough-barked trees, but appear as small discolored depressions, usually with a branch stub in the center. Perennial cankers that are three to four years old exhibit uniform callus ridges that appear targetlike (Fig. 11.7). Small black sterile nodules cover the infected areas (Fig. 11.8). Older cankers are targetlike, elongate [up to 8 feet (2–3 m)], and cause severe distortion of the stem (Fig. 11.9). Sprouts often form below the canker as the crown dies. Once the tree is girdled, black sporodochia, producing large numbers of conidia, develop on the surface of

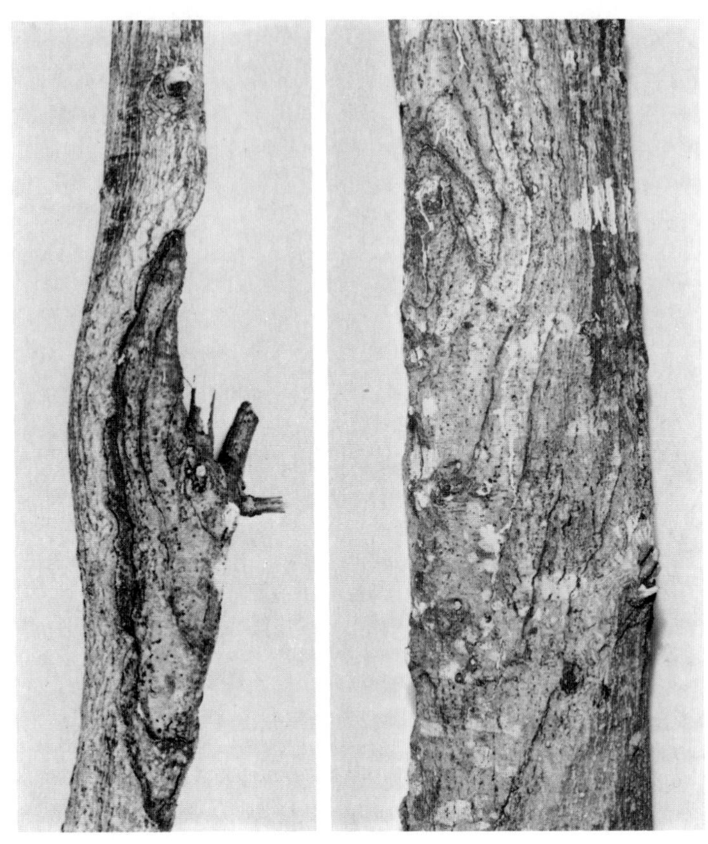

Fig. 11.7 Perennial cankers with target-like callus ridges on red oak infected with *Strumella coryneoidea*.

the dead twigs and stems. Infected logs that have come in contact with wet soil will yield apothecia of the sexual state [*Urnula craterium* (Schw.) Fr.] of *S. coryneoidea.*

Etiology: Method of host penetration is not clear, but presumably the fungus enters through small branches, since branch stubs are usually found at the center of developing cankers (Fig. 11.10). Mycelial fans grow beneath the bark and spread out from the point of entry. The host offers some resistance by forming callus ridges at the canker margins. However, when the tree is dormant, the fungus invades and kills the callus tissue and some additional peripheral tissues. Alternate formation and killing of callus tissue produces cankers that are targetlike. While the stem is still living, small black carbonaceous nodules, made up of a mass of interwoven and fused hyphae, are formed on dead bark of the outer canker. As the canker progresses, the stem becomes distorted. The infected stem is eventually girdled and all parts above

Fig. 11.8 Black sterile nodules on the bark surface of a Strumella canker.

Fig. 11.9 Severe stem distortion caused by *Strumella coryneoidea* on red oak. Note sprout formation below cankers.

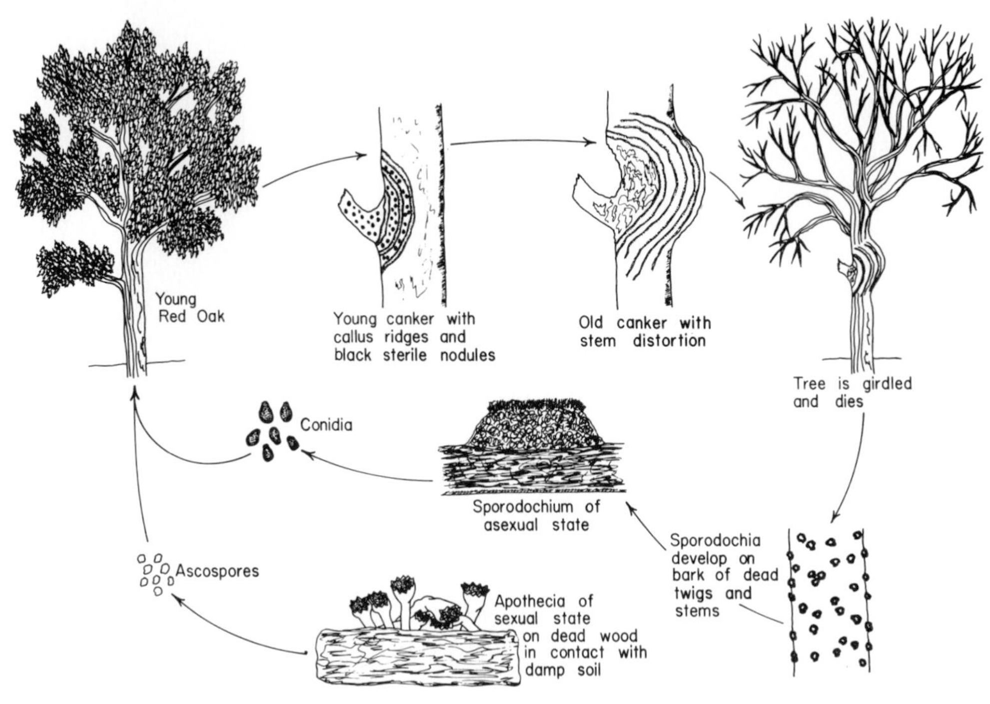

Fig. 11.10 Disease cycle of Strumella canker caused by *Strumella coryneoidea*.

the girdled area die. This usually means the whole tree dies, since *Strumella* cankers are most often found on the lower hole. The infected areas then become covered with sporodochia which produce large numbers of conidia on their surfaces. The conidia are brown and irregularly round to pear-shaped. If the infected tissues fall to the ground and remain in contact with damp soil, apothecia develop and produce hyaline, ellipsoid ascospores. These apothecia are brown to black urn-shaped cups with a wavy margin. Conidia and ascospores are windblown to new hosts.

Control: Control measures are limited to management procedures that favor growth and reproduction of noncankered trees. Felled trees and old dead logs are also a potential source of inoculum of the sexual state of the pathogen, so they should be removed and destroyed.

Selected References

Davidson, R. W. (1950). *Urnula craterium* is possibly the perfect stage of *Strumella coryneoidea*. *Mycologia* **42,** 735–742.

Fergus, C. L. (1951). *Strumella* canker of Bur oak in Pennsylvania. *Phytopathology* **41,** 101–103.

Houston, D. R. (1966). *Strumella* canker of oaks. *U.S. For Ser., For. Pest Leafl.* No. 101.

Sleeth, B., and Lorenz, R. C. (1945). *Strumella* canker of oak. *Phytopathology* **35,** 671–674.

Wolf, F. A. (1958). Mechanism of apothecial opening and ascospore expulsion by the cup-fungus *Urnula craterium*. *Mycologia* **50,** 837–843.

DISEASE: EUTYPELLA CANKER

Primary causal agent: *Eutypella parasitica* Davidson & Lorenz

Hosts: Primarily sugar maple (*Acer saccharum* Marsh.) and red maple (*A. rubrum* L.). Occasionally boxelder (*A. negundo* L.) and Norway maple (*A. platanoides* L.).

Symptomatology: The first observable symptom is the formation of a callus ridge around the portal of entry, usually a branch stub within the first 10 feet (3 meters) of the bole. A slight depression can be seen in the center of the callus area. Older cankers are vertically elongate with firmly attached bark and broad, slightly raised concentric rings of callus tissue (Fig. 11.11). In the center of old cankers perithecia are embedded in the bark with their long necks

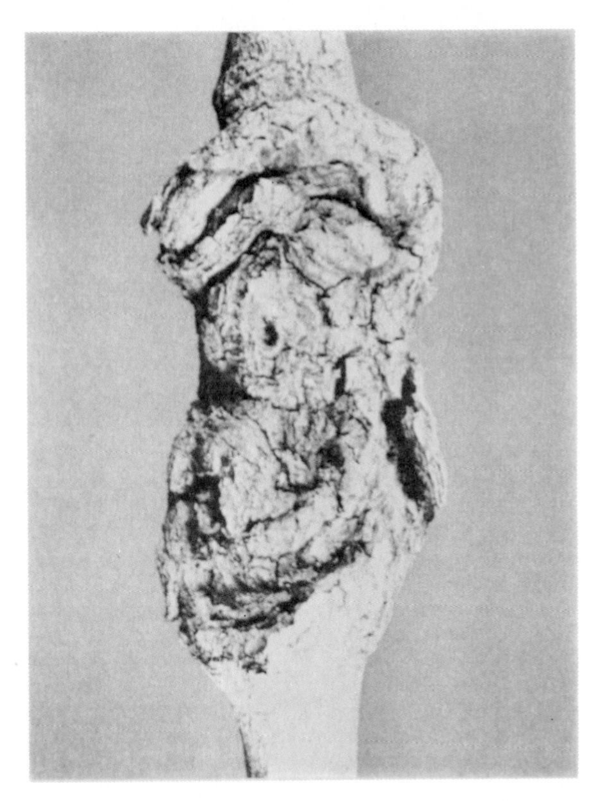

Fig. 11.11 Eutypella canker on young sugar maple stem. (Photograph courtesy of USDA Forest Service.)

Fig. 11.12 Old Eutypella canker on sugar maple stem showing humpback distortion in the cankered area (Photograph courtesy of USDA Forest Service.)

protruding just above the surface. White to tan mycelial mats can be seen beneath the bark at the periphery of the canker. Trees with very old cankers exhibit humpback distortions in the cankered area (Fig. 11.12).

Etiology: Since a dead branch stub is usually present in the center of a canker, initial infection presumably occurs through a branch stub when the tree is young (Fig. 11.13). Artificial inoculations suggest that the fungus must invade the cambium and phloem for infection to occur. Hyphae also penetrate deeply into the sapwood. Alternate growth and killing of callus tissue produces callus ridges. The fungus moves primarily in the phloem, which may explain why the canker advances much faster in a vertical direction. Trees less than three inches (8 cm) in diameter usually are girdled and die within two years. Larger trees are not often killed but may break in the weakened cankered region. Six to eight years after infection, sharply delimited and erumpent stromatic fungal

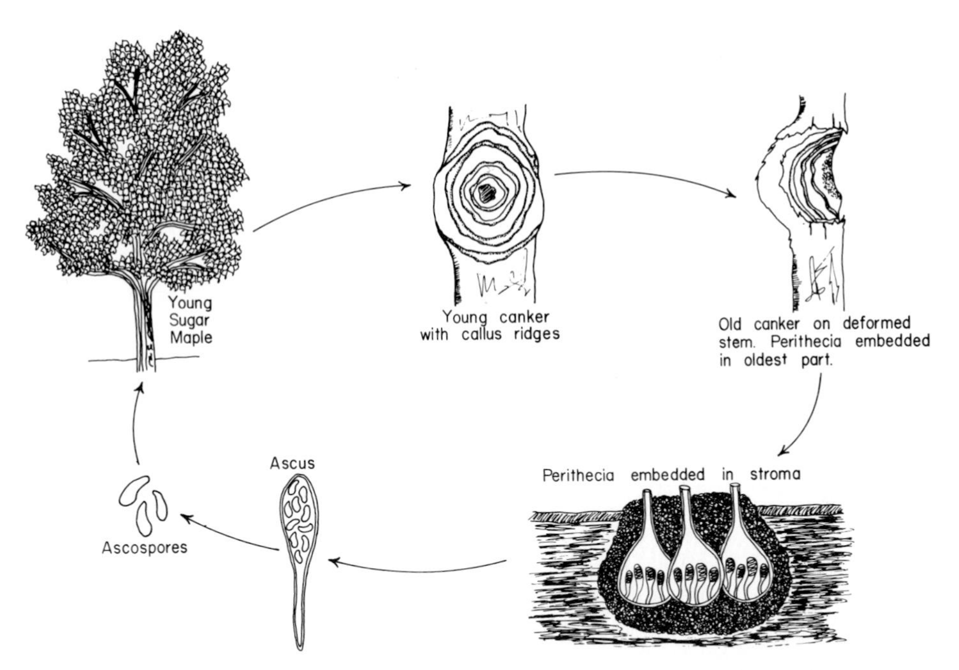

Fig. 11.13 Disease cycle of Eutypella canker caused by *Eutypella parasitica.*

tissues develop in the bark at the center of the canker. Within the stromata, clusters of perithecia develop with long necks that protrude just above the bark surface. Asci produce eight hyaline, slightly curved ascospores. Ascospore release is by forcible discharge and appears to be temperature-humidity triggered. The asexual state of the fungus is known to be formed only in culture.

Control: In forest stands control is not recommended unless the incidence of cankers is high. Then, removal and disposal of infected trees to reduce the inoculum is the prescribed approach. In ornamental trees excising the bark around the canker margins may arrest the spread of the fungus.

Selected References

French, W. J. (1969). *Eutypella* canker of *Acer* in New York. *State Univ. Coll. For. Syracuse Univ., Tech. Publ.* No. 94.

Johnson, D. W., and Kuntz, J. E. (1978). Imperfect stage of *Eutypella parasitica* in culture. *Can. J. Bot.* **56,** 1518–1525.

Johnson, D. W., and Kuntz, J. E. (1979). Eutypella canker of maple: Ascospore discharge and dissemination. *Phytopathology* **69,** 130–135.

Lachance, D., and Kuntz, J. E. (1970). Ascocarp development of *Eutypella parasitica. Can. J. Bot.* **48,** 1977–1979.

Robbins, M. K. (1979). How to identify and minimize damage caused by *Eutypella* canker of maple. *USDA For. Serv., Publ.* **NA-FR-10.**

DISEASE: NECTRIA CANKER

Primary causal agent: *Nectria galligena* Bres.

Hosts: Most hardwoods

Symptomatology: The first indications of the disease are slightly depressed areas of bark around small wounds or dead twigs. In these young perennial cankers the bark may be cracked with callus tissue formed at the edge of the cankered tissue. Older cankers are usually conspicuous, with uniform concentric rings of callus formed to resemble a target (Fig. 11.14). On some hosts (oaks, for example), cankers are irregular in shape. During the fall, the asexual state of

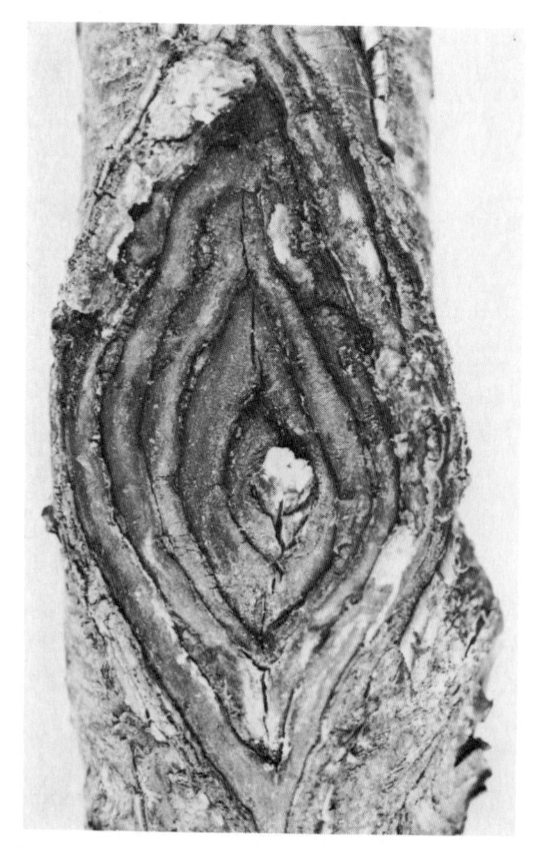

Fig. 11.14 Nectria canker on black birch showing concentric rings of callus in a target shape.

the fungus appears as small white tufts of hyphae (sporodochia) protruding through bark cracks near the periphery of the canker. The following year red, flask-shaped perithecia (sexual state) form singly or in groups on the dead callus tissues.

Etiology: Sexual and asexual spores of the fungus may enter the host through wounds or branch stubs which expose the inner bark or cambium (Fig. 11.15). Once established, the fungus begins to grow outward from the point of infection during the fall and winter. When tree growth resumes in the spring, a layer of callus tissue is formed along the advancing margin or fungal hyphae. In some cases the callus may stop the fungus, but most often the fungus will continue its advance during the dormant season, killing the newly formed callus. Alternate formation and killing of callus produces the targetlike appearance of Nectria cankers. Sporodochia of the asexual state form in the bark cracks along the outer callus ridges. Two types of conidia are produced on the surface of the sporodochia: aseptate microconidia and 1- to 4 or more-septate macroconidia. Conidia are spread to new infection sites by rainsplash and wind. The sexual state of the fungus appears on the same tissues the following year. Minute, red flask-shaped perithecia generally develop in groups on fungal stromata (Fig. 11.16). At maturity the perithecia become dark red and

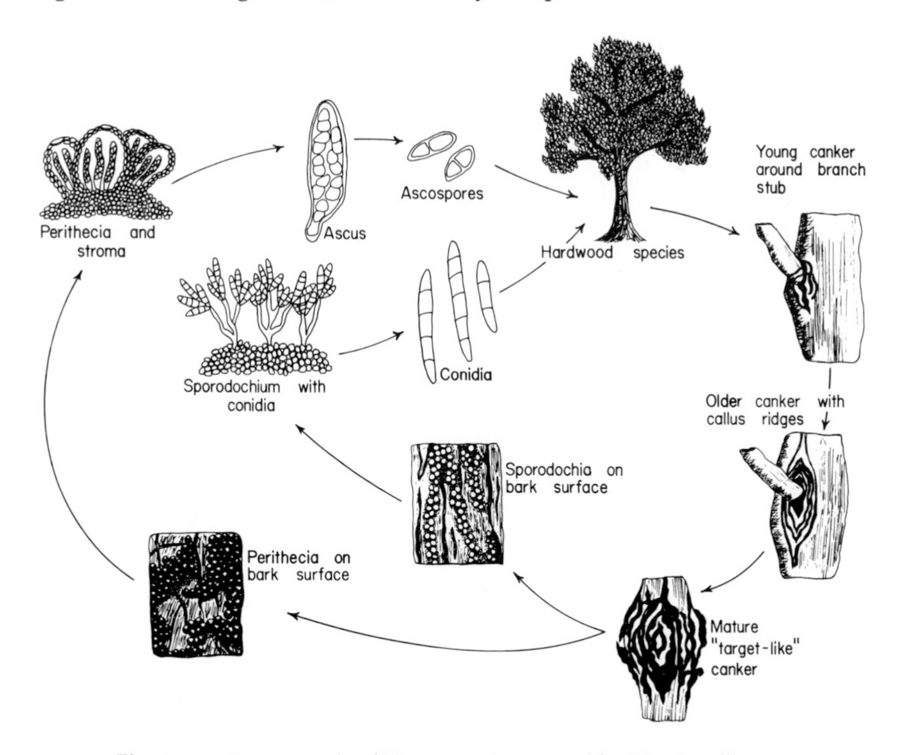

Fig. 11.15 Disease cycle of Nectria canker caused by *Nectria galligena*.

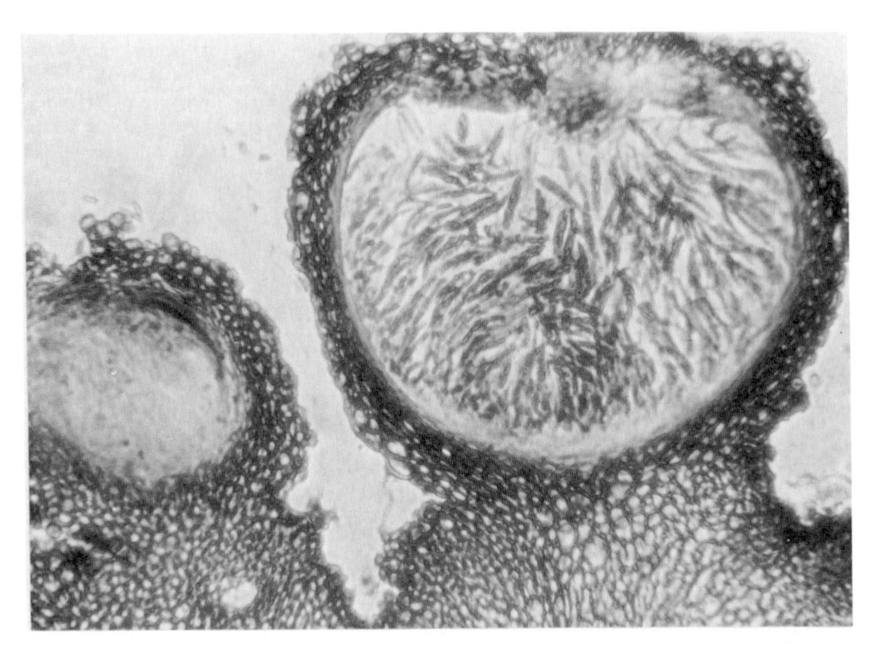

Fig. 11.16 Cross section of perithecia and stroma of *Nectria galligena* showing asci and ascospores. (Photograph from microscope slide prepared by Triarch, Inc., Ripon, Wisconsin.)

the unequally bicellular ascospores are forcibly discharged and dispersed by the wind.

Control: Because the fungus is a common saprophyte, control is difficult. Removal of infected trees and selection for disease-free trees are considered general practices for stand improvement.

Selected References

Brandt, R. W. (1964). Nectria canker of hardwoods. *U.S. For. Serv., For. Pest Leafl.* No. 84.

Grant, T. J., and Spaulding, P. (1939). Avenues of entrance for canker-forming Nectrias of New England hardwoods. *Phytopathology* **29,** 351–358.

Lohman, M. L., and Watson, A. J. (1943). Identity and host relations of Nectria species associated with diseases of hardwoods in the eastern states. *Lloydia* **6,** 77–108.

Lortie, M. (1964). Pathogenesis in cankers caused by *Nectria galligena. Phytopathology* **54,** 261–263.

Spaulding, P. (1952). Nectria canker of hardwoods. *N. Engl. Sect., Soc. Am. For. Tree Pest Leafl.* No. 10.

DISEASE: BEECH BARK DISEASE

Primary causal agent: *Cryptococcus fagisuga* Lind. followed by *Nectria coccinea* var. *faginata* Lohm, Wat. & Ayers.

Fig. 11.17 Beech scale insect on bark of American Beech (A), closeup (B). (Photograph B courtesy of H. Van T. Cotter, University of New Hampshire, Durham.)

Fig. 11.18 American beech with beech bark disease (left) and healthy beech tree (right) (A), closeup of canker (B). (Photographs courtesy of H. Van T. Cotter, University of New Hampshire, Durham.)

Hosts: American beech (*Fagus grandifolia* Ehrh.), European beech (*F. sylvatica* L.) and its varieties

Symptomatology: Evidence of the disease is first seen in the fall as individual white, woolly specks on the trunk and branches (Fig. 11.17). This is *C. fagisuga,* the woolly beech scale. Heavy infestations will make the bark appear as though dusted with snow. Once the fungus becomes established, small circular to horizontally elliptic cankers appear as slightly sunken lesions with cracks in the bark (Fig. 11.18). White sporodochia followed by clusters of red perithecia grow on the cankered areas (Fig. 11.19). Coalescence of cankers results in killing of large areas of bark, sometimes girdling the tree, followed by wilting of the foliage and progressive death of twigs and branches.

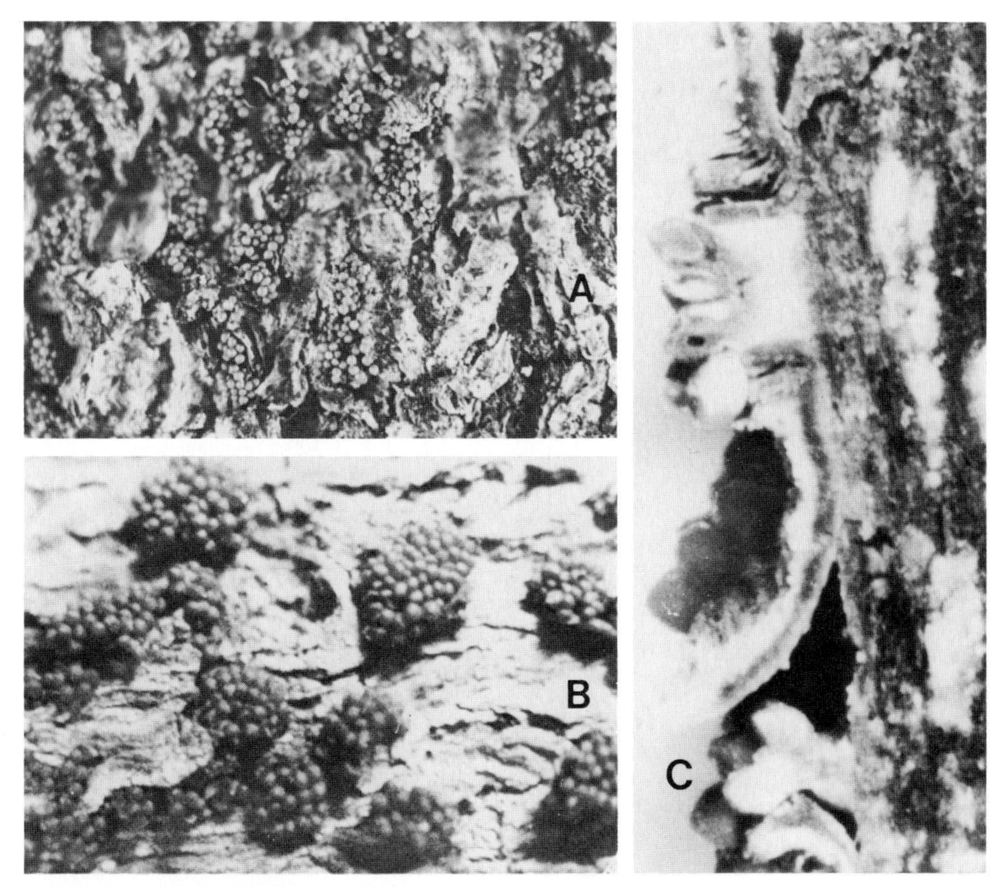

Fig. 11.19 Clusters of perithecia of *Nectria coccinea* var. *faginata* on cankered bark of American beech (A), closeup (B), cross section of bark surface (C). (Photograph C courtesy of H. Van T. Cotter, University of New Hampshire, Durham.)

Etiology: Minute yellow larvae of the scale insect hatch in late summer (Fig. 11.20). These comprise the "crawler" stage. By early fall they insert their sucking organ (stylet) into the living bark, moult, and secrete a white, wool-like wax coat around themselves. The insects overwinter in this stage (second instar larvae) and complete development by moulting again to become adults the following spring. During early summer, eggs are laid under the woolly coat of the adult insects. When several insects colonize a small area, groups of host cells die, leaving minute cracks in the bark. The fungus presumably enters through these cracks. Infections of *N. coccinea* var. *faginata* always seem to be associated spatially with the earlier infestation of *C. fagisuga,* but the presence of *C. fagisuga* does not necessarily mean infection of the fungus will occur. If infections do occur and are isolated, small circular cankers with cracked bark develop. The asexual state of the fungus, *Cylindrocarpon faginata* Booth, appears as clusters of white cushion-shaped sporodochia that push out from the bark cracks. Two types of conidia are produced on the surface of the sporodochia: aseptate microconidia and 1- to 4 or more-septate macroconidia. During the fall clusters of perithecia develop on yellowish stromata which push through the outer bark. The perithecia are dark to medium red and can be seen with the naked eye. This is the sexual state of the fungus. During late winter and early spring, unequally bicellular ascospores are released and spread by wind, rain, and probably to a lesser extent by insects and birds. The

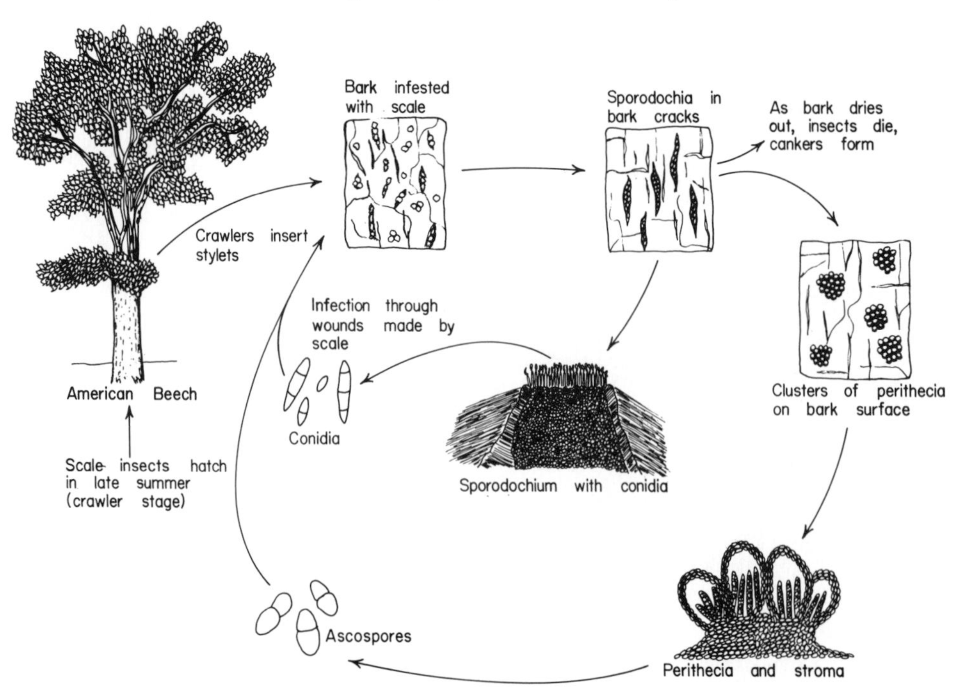

Fig. 11.20 Disease cycle of beech bark disease caused by *Nectria coccinea* var. *faginata*.

relative extent to which conidia and ascospores contribute to new infections is not known.

Control: Once the fungus becomes established in a tree, no known control is available. On ornamental trees, the disease has been successfully controlled by applying chemical insecticides to control the insect. In the forest, control is more difficult and at present is limited to certain silvicultural practices of questionable effectiveness. The extent to which natural factors influence control is not clear, but some trees in heavily infected areas appear resistant. Further, the twice-stabbed lady beetle (*Chilocorus stigma* Say) actively feeds on the scale insect; the fungus *Gonatorrhodiella highlei* A. L. Sm. is parasitic on *N. coccinea* var. *faginata,* and in Europe the scale insect is excluded from areas of bark colonized by the fungus *Ascodichaena rugosa* Butin.

Selected References

Cotter, H. V. T. (1977). Beech bark disease: Fungi and other associated organisms. M. S. Thesis, Univ. of New Hampshire, Durham.

Ehrlich, J. (1934). The beech bark disease, a *Nectria* disease of *Fagus*, following *Cryptococcus fagi* (Baer.). *Can. J. Res.* **10,** 593–692.

Houston, D. R. (1975). Beech bark disease: The aftermath forests are structured for a new outbreak. *J. For.* **73,** 660–663.

Houston, D. R., Parker, E. J., and Lonsdale, D. (1979). Beech bark disease: Patterns of spread and development of the initiating agent, *Cryptococcus fagisuga. Can. J. For. Res.* **9,** 336–344.

Houston, D. R., Parker, E. J., Perrin, R., and Lang, K. J. (1979). Beech bark disease: A comparison of the disease in North America, Great Britain, France and Germany. *Eur. J. For. Pathol.* **9,** 199–211.

Lonsdale, D. (1979). Beech bark disease: One disorder or several? *Ecologist* **9,** 136–138.

Parker, E. J. (1974). Beech bark disease. Ph.D. Thesis, Univ. of Surrey, Guilford, England.

DISEASE: HYPOXYLON CANKER

Primary causal agent: *Hypoxylon mammatum* (Wahl.) Mill.

Hosts: Primarily quaking aspen (*Populus tremuloides* Michx.), occasionally other *Populus* spp.

Symptomatology: Slightly sunken, yellow-orange areas with irregular margins appear on the bark, often around a branch stub near the base of the tree. The discolored areas increase in size and the outer bark (periderm) in the oldest part of the canker becomes blistered (Fig. 11.21). Removal of the blistered bark will reveal conidial pillars (coremia) (Fig. 11.22). Older cankers will have clusters of perithecia immersed in stromata arising on the site of the conidial pillars (Fig. 11.23). The grayish stromata surrounding black perithecial heads are diagnostic. Wind breakage in the area of cankers is common (Fig. 11.24).

Etiology: Method of entry of *H. mammatum* into the host is not certain, but most cankers are associated with branch stubs and it is assumed that such stubs

Fig. 11.21 Hypoxylon canker associated with a dead branch on aspen. (Photograph courtesy of Alex L. Shigo, USDA Forest Service.)

Fig. 11.22 Conidial pillars of *Hypoxylon mammatum* beneath bark of aspen. (Photograph courtesy of USDA Forest Service.)

Fig. 11.23 Clusters of perithecia of *Hypoxylon mammatum* with grayish stromata surrounding black perithecial heads.

Fig. 11.24 Wind breakage in area of Hypoxylon canker on aspen stem. (Photograph courtesy of Botany and Plant Pathology Dept., University of New Hampshire, Durham.)

may be the site of entry (Fig. 11.25). Recent evidence also suggests that certain insects may be involved either as vectors or as wounding agents. As the pathogen becomes established it invades all bark zones except the periderm. Within two years the periderm becomes blistered and sloughs off as a result of the developing conidial pillars beneath. Hyaline, generally ellipsoid conidia develop on the pillars. The following year, grayish stromata develop on the site of the conidial pillars. Five to ten perithecia with projecting, black, ostiolate necks are immersed in each stroma. Brown, ellipsoid ascospores with a longitudinal furrow are forcibly discharged from perithecia and dispersed by wind to other susceptible hosts. Conidia are considered to have little significance in the transmission of the fungus.

Control: No direct control procedures are known. However, disease occurrence has been shown to be reduced in dense stands with a closed canopy, suggesting that thinning should be used with care as a management practice. The

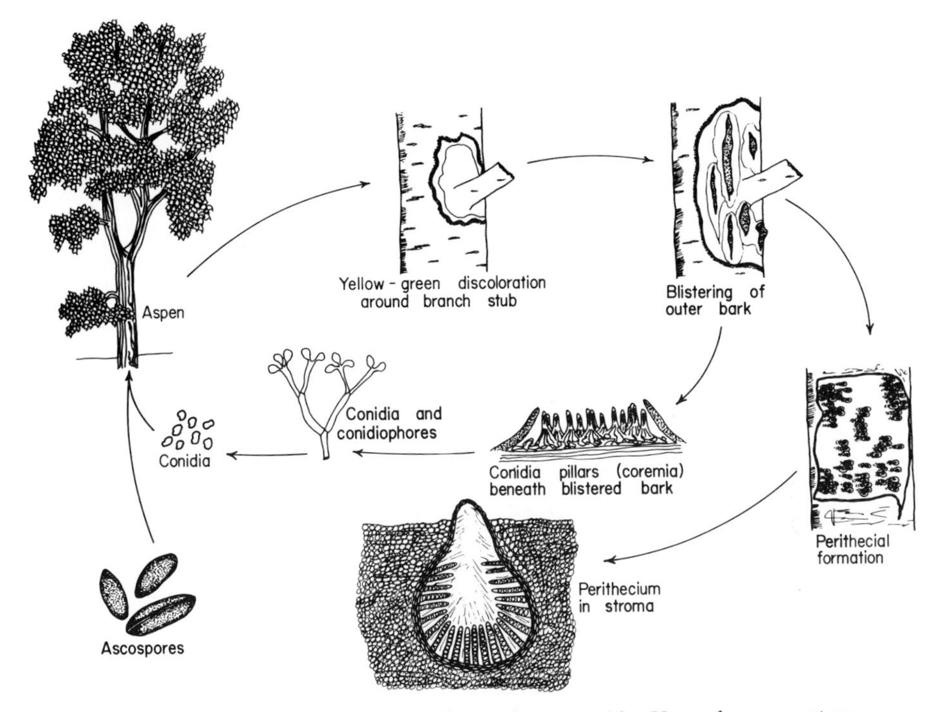

Fig. 11.25 Disease cycle of Hypoxylon canker caused by *Hypoxylon mammatum*.

most promising efforts for control of the disease appear to be in breeding programs for development of resistant strains.

Selected References

Anderson, R. L., and Anderson, G. W. (1968). Relationship between density of quaking aspen and incidence of Hypoxylon canker. *For. Sci.* **14,** 107–112.

Anderson, R. L., and Anderson, G. W. (1979). Hypoxylon canker of aspen. *U.S. For. Serv., For. Pest Leafl.* No. 6 (revised).

Bagga, D. K., and Smalley, E. B. (1974). The development of Hypoxylon canker of *Populus tremuloides:* Role of ascospores, conidia, and toxins. *Phytopathology* **64,** 654–658.

Berbee, J. G., and Rogers, J. D. (1964). Life cycle and host range of *Hypoxylon pruinatum* and its pathogenesis on poplars. *Phytopathology* **54,** 257–261.

French, D. W., Hodges, C. S., Jr., and Froyd, J. D. (1969). Pathogenicity and taxonomy of *Hypoxylon mammatum. Can. J. Bot.* **47,** 223–226.

Manion, P. D. (1975). Two infection sites of *Hypoxylon mammatum* in trembling aspen (*Populus tremuloides*). *Can. J. Bot.* **53,** 2621–2624.

DISEASE: SCLERODERRIS CANKER

Primary causal agent: Gremmeniella abietina (Lagerb.) Morelet. Two strains are known to exist in North America:

1. North American strain—produces conidia and ascospores.
2. European strain—Produces conidia and rarely, ascospores.

Intermediate strains (potentially natural hybrids) of these two strains are known and are under investigation.

Hosts: Primarily Scotch pine (*Pinus sylvestris* L.), red pine (*P. resinosa* Ait.), and jack pine (*P. banksiana* Lamb.). Over 40 species or varieties of conifers (primarily pine and spruce) have shown various degrees of susceptibility.

Symptomatology (on red pine): First evidence of the disease is seen in the spring when previous year's needles begin to turn yellow and then red from the base to the tip. Terminal buds of infected shoots do not develop. During the summer, needles on infected shoots are cast, leaving bare shoots that look like developing candles (Fig. 11.26). Cankerlike lesions can often be seen on the exposed shoots. Woody tissue beneath the bark in the transition area between living and dead tissues will exhibit a yellow-green discoloration, most easily seen in summer. In late fall or spring, pycnidia form in leaf scars near the transition area (Fig. 11.27). During the fall or the following spring, the North American strain also produces small (1 mm diam.) apothecia on infected shoots (Fig. 11.28). Generally, only young trees are affected, but older trees have been found to be affected (Fig. 11.29).

Etiology: Infection occurs on newly developing shoots of pine during shoot elongation in the spring (Fig. 11.30). Method of penetration is not known. The fungus remains in the invaded tissues during the summer of year 1. No

Fig. 11.26 Needle cast of shoots of red pine infected with *Gremmeniella abietina*. (Photograph courtesy of Dale R. Bergdahl, University of Vermont, Burlington.)

Fig. 11.27 Pycnidia of *Gremmeniella abietina* in leaf scars of young pine twigs. (Photograph courtesy of Dale R. Bergdahl, University of Vermont, Burlington.)

observable symptoms are present at this time. During the fall and winter of year 1, the fungus colonizes the infected shoot. During the spring of year 2 cankerlike lesions occur on the infected shoots. The needles turn yellow and then red at their bases and discolor progressively toward the tips. Needles are cast during the summer as dieback of woody tissues progresses. Pycnidia develop during this time (they may occur earlier), but conidia are not abundant until late fall or the following spring (year 3). The conidia are hyaline, cylindrical, somewhat curved, and mostly 3-septate. Spore dissemination is primarily by rainsplash. The European strain seldom produces a sexual state, but the North American strain does, with apothecia forming in the fall of year 2 and the spring of year 3. Ascospores are hyaline, ellipsoid, slightly curved, and 3-septate, and are dispersed during July and August. Seriously infected trees eventually die.

Fig. 11.28 Apothecia of *Gremmeniella abietina* on pine. (Photograph courtesy of USDA Forest Service.)

Fig. 11.29 Red pine stand in New York State heavily infected with *Gremmeniella abietina*. (Photograph courtesy of Dale R. Bergdahl, University of Vermont, Burlington.)

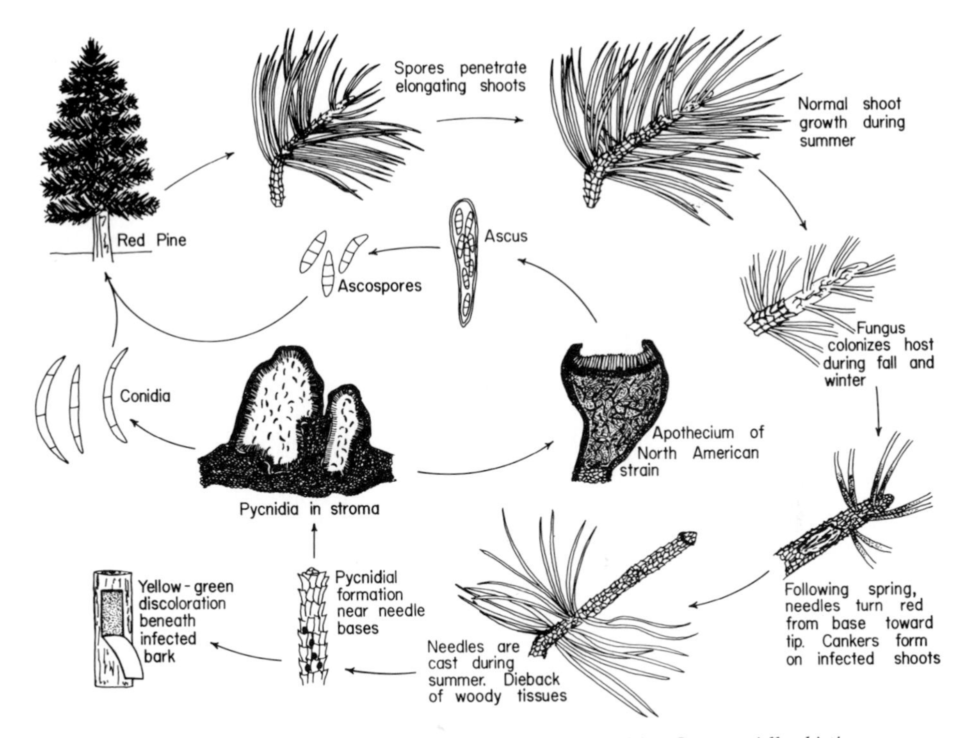

Fig. 11.30 Disease cycle of Scleroderris canker caused by *Gremmeniella abietina.*

Control: In nurseries fungicides have been used with varying success. Fewer infections have been reported in red pine plantations when silvicultural practices of pruning and thinning were used. In areas of high risk (mortality), planting of resistant or immune species is recommended.

Selected References

Dorworth, C. E. (1971). Diseases of conifers incited by *Scleroderris lagerbergii* Gremmen: A review and analysis. *Can. For. Serv., Publ.* No. 1289.

Magasi, L. P. (1979). Should Scleroderris scare us? *Can. For Serv., Marit. For. Res. Cent. Inf. Rep.* **M-X-100.**

Nicholls, T. H. (1979). Scleroderris canker in conifers. *Am. Christmas Tree J.* **23,** 23–26.

Skilling, D. D. (1971). Epidemiology of *Scleroderris lagerbergii. Eur. J. For. Pathol.* **2,** 16–21.

Skilling, D. D. (1974). Fungicides for control of Scleroderris canker. *Plant Dis. Rep.* **58,** 1097–1100.

Skilling, D. D., and O'Brien, J. T. (1973). How to identify Scleroderris canker and red pine shoot blight. *USDA For. Serv., Publ.* **761-930.**

DISEASE: CYTOSPORA CANKER

Primary causal agents: Cytospora chrysosperma (Pers.) Fr. on hardwoods, *C. kunzei* Sacc. on conifers

Hosts: More than 70 species of hardwoods and conifers. Most common on Colorado blue spruce (*Picea pungens* Engelm.), Norway spruce [*P. abies* (L.) Karst.], maples (*Acer* spp.), mountain ash (*Sorbus americana* Marsh.), poplars (*Populus* spp.), and willows (*Salix* spp.).

Symptomatology: On hardwoods, sunken elongate cankers appear around branch stubs or wounds on the trunk or branches. Cankers are usually discolored and surrounded by a callus ridge. The fungus appears on dead bark as small pimplelike protrusions. These are pycnidia. During wet weather the pycnidia produce spores in cirrhi. On conifers, specifically blue spruce, dying back or flagging of branches from the lower portion of the tree toward the top is the most noticeable symptom (Fig. 11.31). Needles turn brown and may remain attached for 1–2 years. Cankers develop around branch stubs or wounds and exhibit considerable amounts of resin (Fig. 11.32). Pycnidia with cirrhi can often be found on the surface of cankered tissue.

Etiology: Conidia of the fungus enter the tree presumably through branch stubs or wounds (Fig. 11.33). The fungus colonizes and kills the cambium and inner bark. Growth proceeds faster along the longitudinal axis of the infected branch than around it, thus producing elongated cankers. Small branches may be killed within 1–2 years, whereas larger branches and the trunk may not be girdled for several years. As the bark dies, it becomes cracked, producing large amounts of resin. Multiloculate pycnidia develop in the dead tissue and exude

Fig. 11.31 Colorado blue spruce infected with *Cytospora kunzei* (right) and healthy spruce (left). Note lower branch dieback on infected tree. (Photograph couresy of Avery R. Rich, University of New Hampshire, Durham.)

Fig. 11.32 Resin flow from Cytospora canker on spruce branch.

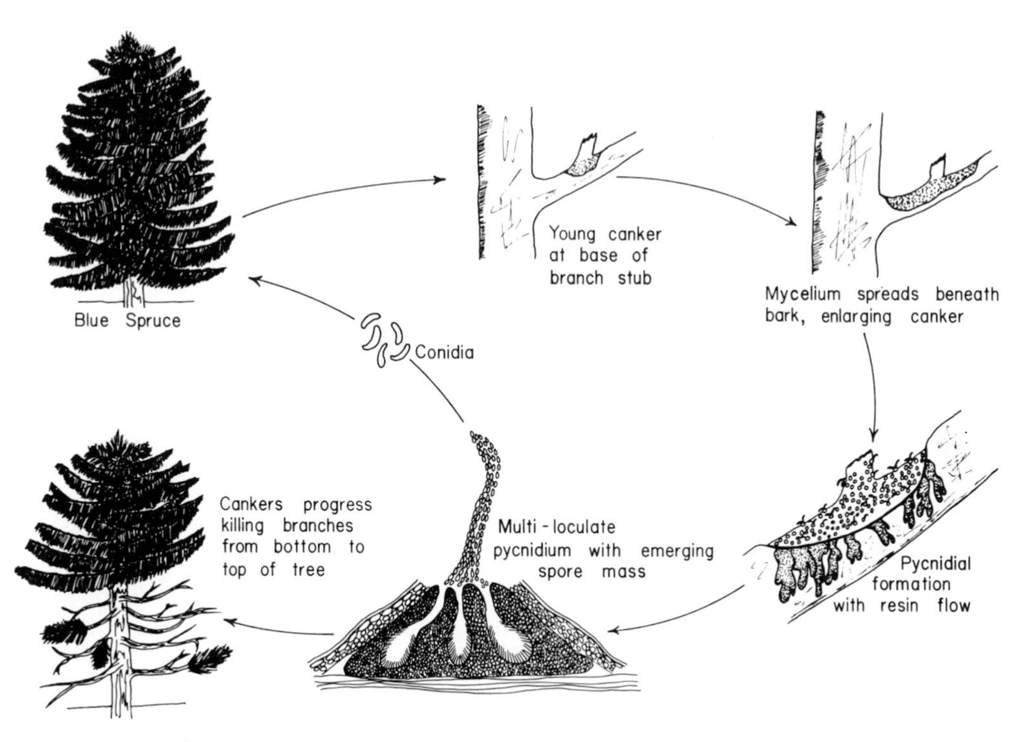

Fig. 11.33 Disease cycle of Cytospora canker caused by *Cytospora kunzei*.

hyaline, elongate-curved conidia in a gelatinous matrix resembling a coiled thread (cirrhus). Rainsplash washes the conidia to other infection sites, or they may be transported by insects or pruning tools. Perithecia of the perfect state (*Valsa* sp.) are seldom found.

Control: Trees like Colorado blue spruce are susceptible primarily when they are planted at the outer limits of their natural range. Since the fungus is considered a weak parasite, attempts should be made to keep susceptible species vigorous by careful site selection, fertilization, and watering. Additional control can be achieved by pruning and destroying all infected branches during dry weather. Also, the sterilization of pruning tools is necessary after each cut to prevent spread of the pathogen.

Selected References

Bloomberg, W. J. (1962). Cytospora canker of poplars: Factors influencing the development of the disease. *Can. J. Bot.* **40,** 1271–1280.

Christensen, C. M. (1940). Studies on the biology of *Valsa sordida* and *Cytospora chrysosperma. Phytopathology* **30,** 459–475.

French, D. W. (1961). Cytospora canker in Minnesota. *Proc. Int. Shade Tree Conf.* **37,** 126–128.

Partridge, A. D. (1966). Tip dieback of young grand fir caused by Cytospora canker. *Univ. Idaho, Stn. Note* No. 5.

Scharpf, R. F., and Bynum, H. H. (1975). Cytospora canker of true firs. *U.S. For. Serv., For. Pest Leafl.* No. **146.**

12

Rust Diseases

INTRODUCTION

Rust diseases are caused by basidiomycetous fungi in the order Uredinales. Members of this group are known for the production of several spore types and often require two hosts (heteroecious) to complete their life cycles. Rust fungi can have up to five spore stages in their life cycles which occur in the following order: basidiospores, pycniospores (spermatia), aeciospores, urediospores, and teliospores. Basidiospores, aeciospores, and urediospores are infective and are able to enter plant tissues, whether by direct penetration or by growing through a stomate. Pycniospores and teliospores have a sexual function. Pycniospores are responsible for the fusion (plasmogamy) of two monokaryotic (single nucleus) rust strains to form dikaryotic (two nuclei) hyphae. This process is known as spermatization. In the teliospore stage the nuclei fuse (karyogamy) and undergo a reduction division (meiosis) to produce monokaryotic basidiospores. If a rust fungus possesses all five spore stages it is termed macrocyclic; if it lacks urediospores, it is termed demicyclic; if it lacks both aeciospores and urediospores, it is termed microcyclic. All rust fungi must contain basidiospores, pycniospores, and teliospores.

Rust diseases of trees can be placed into three general categories based on symptomatology on the tree host: (1) canker rusts, (2) gall rusts, and (3) leaf rusts. Examples of each type, white pine blister rust (canker), pine oak gall rust and fusiform rust (gall), cedar-apple rust (gall and leaf), and ash rust (leaf), will be examined in detail.

Selected References

Bingham, R. I., Hoff, R. J., and McDonald, G. I., eds. (1972). Biology of rust resistance in forest trees: NATO-IUFRO Advanced Study Proceedings. *Misc. Publ.—U.S. Dep. Agric.* No. 1221.

Littlefield, L. J., and Heath, M. C. (1979). "Ultrastructure of Rust Fungi." Academic Press, New York.

Ziller, W. G. (1974). The tree rusts of Western Canada. *Can. For. Serv., Publ.* No. 1329.

DISEASE: WHITE PINE BLISTER RUST

Primary causal agent: Cronartium ribicola Fisch.

Hosts: 5-needled *Pinus* spp. (white pines), *Ribes* spp. (currants and gooseberries)

Fig. 12.1 Abundant resin flow on white pine from white pine blister rust canker. (Photograph courtesy of Avery R. Rich, University of New Hampshire, Durham.)

Fig. 12.2 Cracked bark on canker face of white pine infected with *Cronartium ribicola.*

Symptomatology: Yellow-brown cankers appear on infected pine branches. Cankers girdle branches and advance toward the main stem and eventually girdle it. Abundant resin flows from the canker margin and coats the trunk and lower branches (Fig. 12.1). The crown above the canker exhibits sparse growth, becomes progressively chlorotic and eventually dies. Occasionally the trunk may break off at the canker. The bark on the canker face is cracked (Fig. 12.2). In the spring pycnial ooze flows from it initially, followed later by blister formation on the edge of the canker. Under the blisters the white peridial membranes first appear and later rupture to reveal yellow-orange masses of aeciospores (Fig. 12.3). Black pycnial scars can be seen with the naked eye around the periphery of the cankers (Fig. 12.4). On *Ribes* orange uredial pustules form mainly on the undersides of leaves in the summer (Fig. 12.5). These give rise to telial horns in late summer and early fall (Fig. 12.6). Heavy infections can lead to premature leaf drop.

Fig. 12.3 Aecial blisters on canker face of white pine infected with *Cronartium ribicola.* (Photograph courtesy of Avery R. Rich, University of New Hampshire, Durham.)

Etiology: The pathogen is a macrocyclic rust with five spore stages on two hosts (Fig. 12.7). In the fall germinating teliospores on *Ribes* leaves produce basidiospores which are windblown to susceptible white pines. Basidiospores enter the needles through the stomates. The point of infection is marked by a pale chlorotic spot. The fungus invades progressively into the twigs, branches, and often into the main stem. Two–three years after initial infection, cankers are formed and develop over the next several years. Pycnia are first produced under the bark (Fig. 12.8) and the next year following spermatization of the rust, aecia are produced and continue to be produced each year as long as the canker is active. Aeciospores (Fig. 12.9) are windblown to leaves of *Ribes* where they penetrate the leaves and produce uredia, which form urediospores that can reinfect other leaves. Telia replace the uredia on the *Ribes* leaves later in the season.

Fig. 12.4 Black pycnial scars on canker face of white pine infected with *Cronartium ribicola.*

Control: Traditional control procedures include removal of *Ribes* plants within 900 feet (300 m) of susceptible white pines. However, recent evidence suggests much longer transport of basidiospores from *Ribes* to pine in some areas, due to nighttime air drainage and recirculation patterns. Avoid planting white pines in these high hazard zones. Pruning of lower branches will not only remove existing branch cankers, but may reduce the likelihood of their occurrence.

Selected References

Anderson, R. L. (1973). A summary of white pine blister rust research in the Lake States. *U.S. For. Serv., Gen. Tech. Rep. NC* **NC-6.**

Colley, R. H. (1918). Parasitism, morphology and cytology of *Cronartium ribicola. J. Agric. Res.* **15,** 619–660.

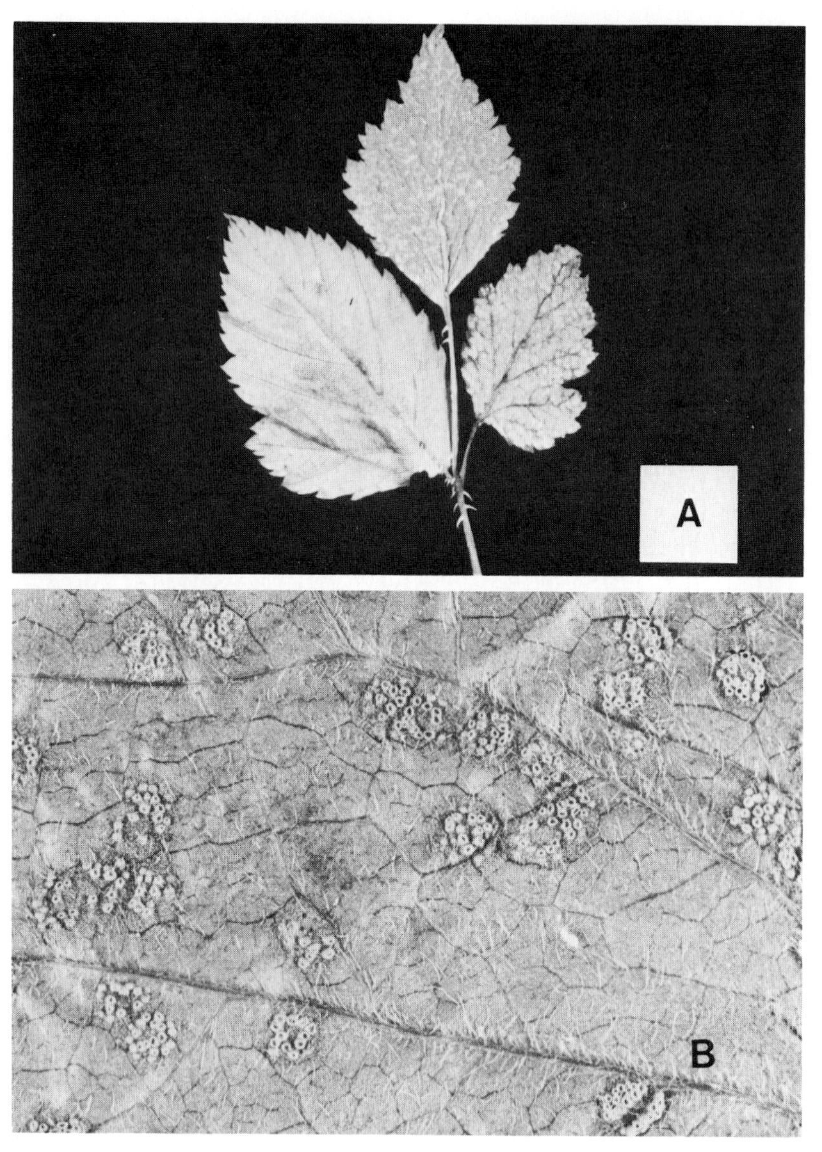

Fig. 12.5 Uredial pustules on the undersides of *Ribes* leaves infected with *Cronartium ribicola* (A), closeup (B). (Photograph A courtesy of William E. MacHardy, University of New Hampshire, Durham.)

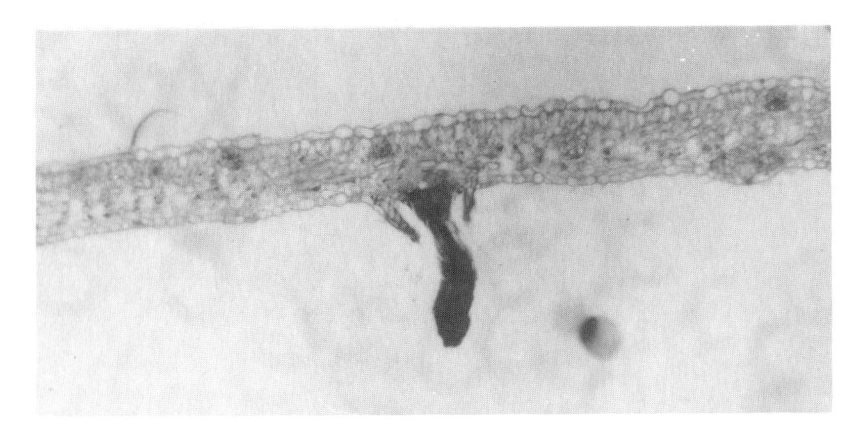

Fig. 12.6 Cross section through telial horn of *Cronartium ribicola* on *Ribes* leaf. 80×. (Photograph from microscope slide prepared by Carolina Biological Supply Company, Burlington, North Carolina.)

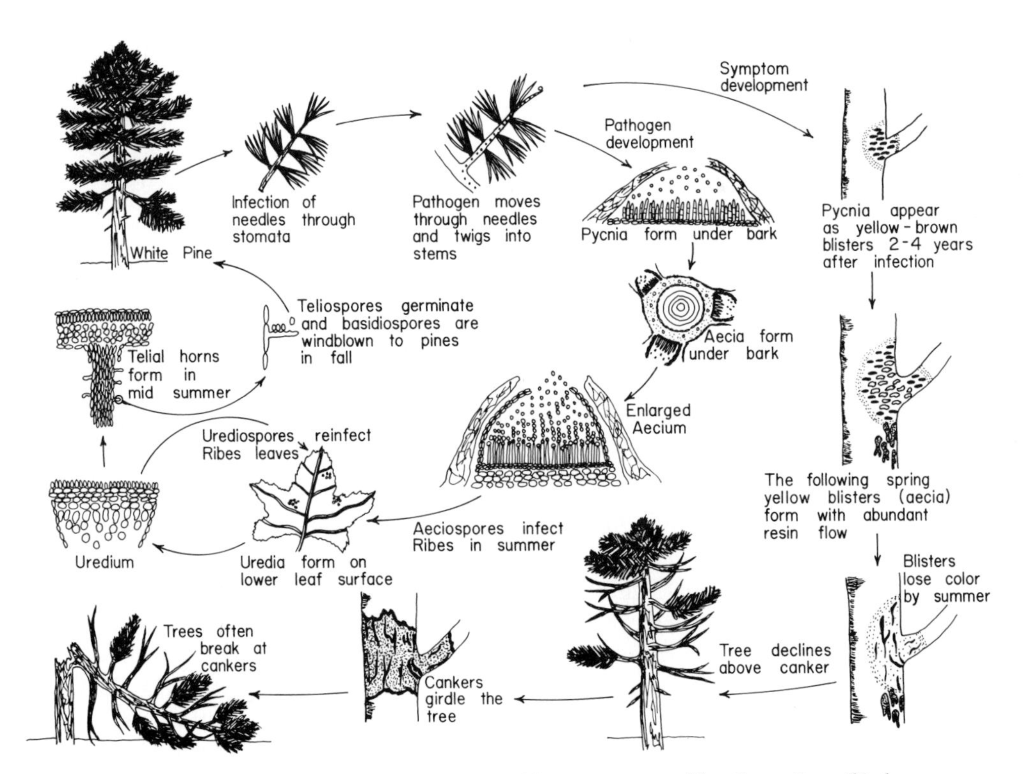

Fig. 12.7 Disease cycle of white pine blister rust caused by *Cronartium ribicola.*

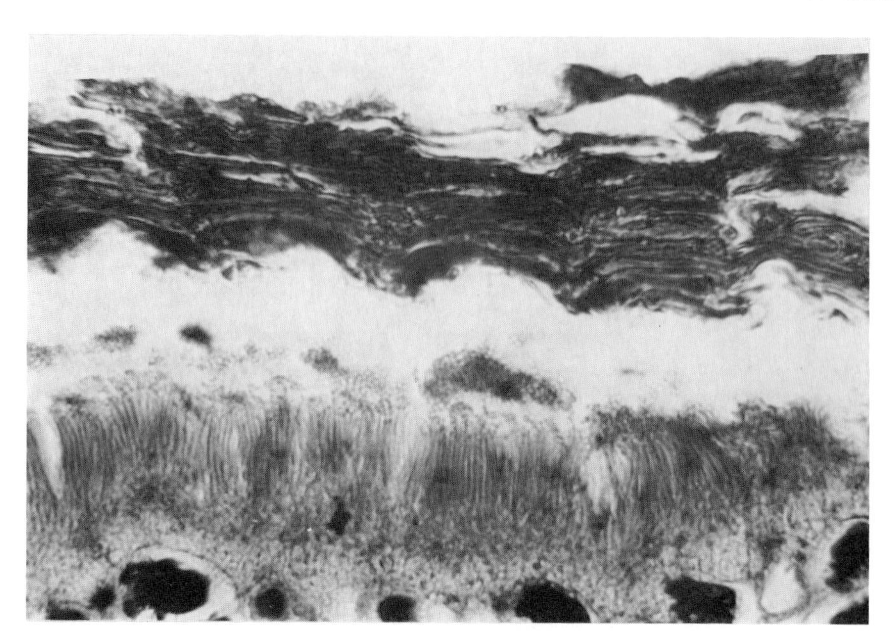

Fig. 12.8 Portion of a pycnium beneath the bark of white pine infected with *Cronartium ribicola* (Photograph from microscope slide prepared by Carolina Biological Supply Company, Burlington, North Carolina.)

Peterson, G. W., and Jewell, F. F. (1968). Status of American stem rust of pine. *Annu. Rev. Phytopathol.* **6,** 23–40.

DISEASE: PINE-OAK GALL RUST (EASTERN GALL RUST)

Primary causal agent: *Cronartium quercuum* (Berk.) Miy. ex Shirai

Hosts: Many 2- and 3-needled *Pinus* spp. (hard pines), *Quercus* spp. (primarily red oak group)

Symptomatology: Globose, spherical galls are produced on the branches and/or stems of susceptible pines. Galls completely encircle small branches while large branches may be only surrounded partially (Fig. 12.10). Dieback above the gall and wind breakage at the gall are common. Trees of all sizes from seedlings to mature trees can be infected, but the disease is most serious on seedlings, which often are killed. In the spring galls are covered with white spore sacs which soon rupture to release red-orange aeciospores (Fig. 12.11). These spores are windblown to the expanding oak leaves. After infection, small uredial lesions with yellow-orange urediospores are formed on the

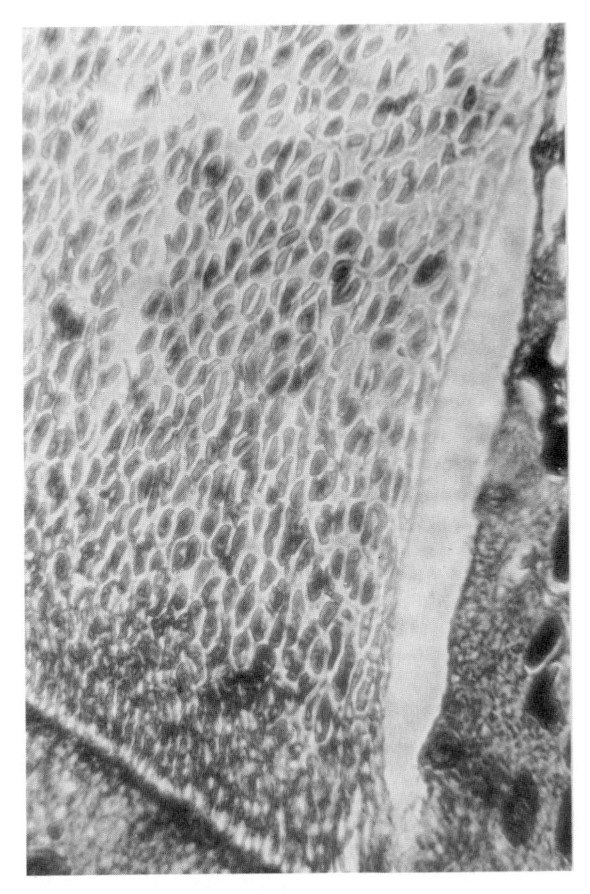

Fig. 12.9 Aeciospores of *Cronartium ribicola* to left of aecial margin. (Photograph from microscope slide prepared by Carolina Biological Supply Company, Burlington, North Carolina.)

lower leaf surface. Later in the season uredia are replaced by short, brown telial columns (Fig. 12.12) which remain on the leaves over winter.

Etiology: The pathogen is a macrocyclic rust with five spore stages on two hosts (Fig. 12.13). Basidiospores infect needles through stomates of susceptible hard pines and invade shoot and stem tissues. A woody gall is induced, and usually within a year following infection, pycnia are formed. After successful spermatization between compatible mating types of the fungus, aecia are produced the following season. Aecia and pycnia will continue to be produced in alternate years as long as the gall is active. Blue stain fungi and decay fungi frequently invade older galls. Aeciospores are windblown to expanding oak leaves where they germinate and penetrate through stomates. Uredia are formed on the lower leaf surface and urediospores are produced. Additional

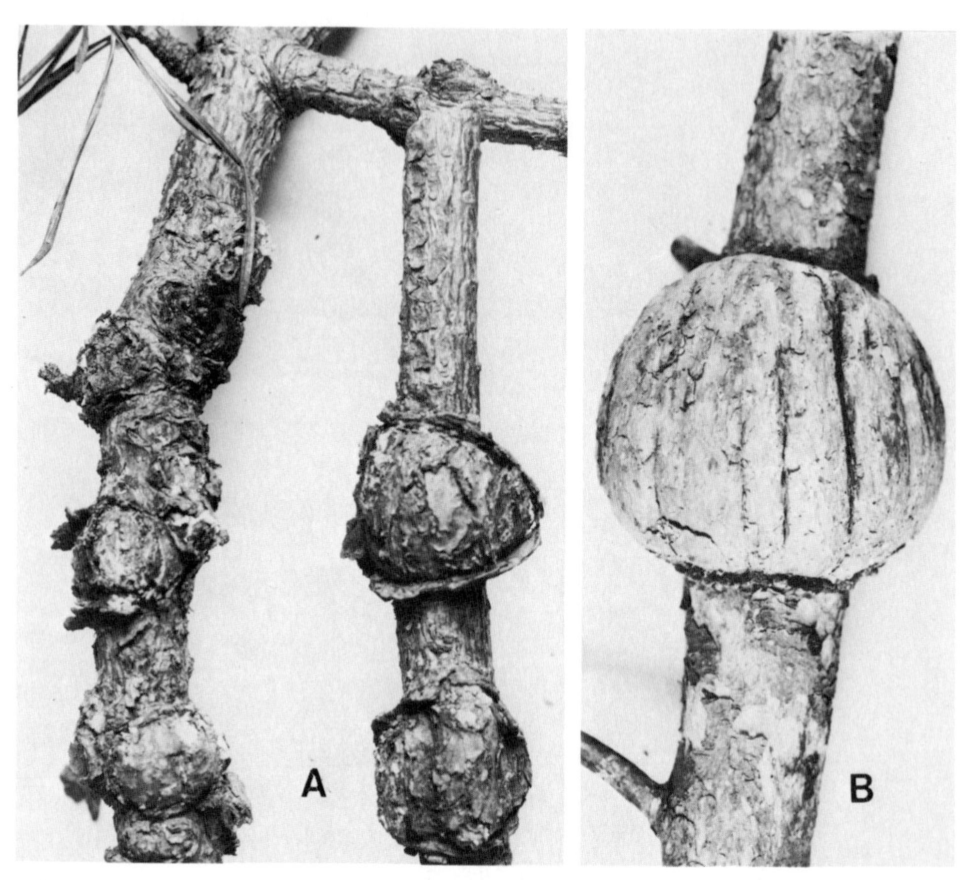

Fig. 12.10 Galls on pine stems induced by *Cronartium quercuum* that partially (A) and completely (B) surround stems. (Photograph A courtesy of Shade Tree Laboratories, University of Massachusetts, Amherst.)

oak infection by urediospores can occur until the leaves are fully formed, but then the leaves become resistant. Telia replace uredia on the leaves later in the summer. Teliospores mature over winter and germinate to produce basidiospores the following spring.

Control: Infected seedlings should be removed as soon as symptoms are noticed. Avoid planting susceptible pines near oaks or wherever pine-oak gall rust is common. Protectant fungicides should be applied to susceptible pine seedlings during the spring infection period.

Selected References

Anderson, N. A. (1963). Eastern gall rust. *U.S. For. Serv., For. Pest Leafl.* No. 80.

Anderson, G. V., and French, D. W. (1965). Differentiation of *Cronartium guercuum* and *Cronartium coleosporiodes* on the basis of aeciospore germ tubes. *Phytopathology* **55,** 171–173.

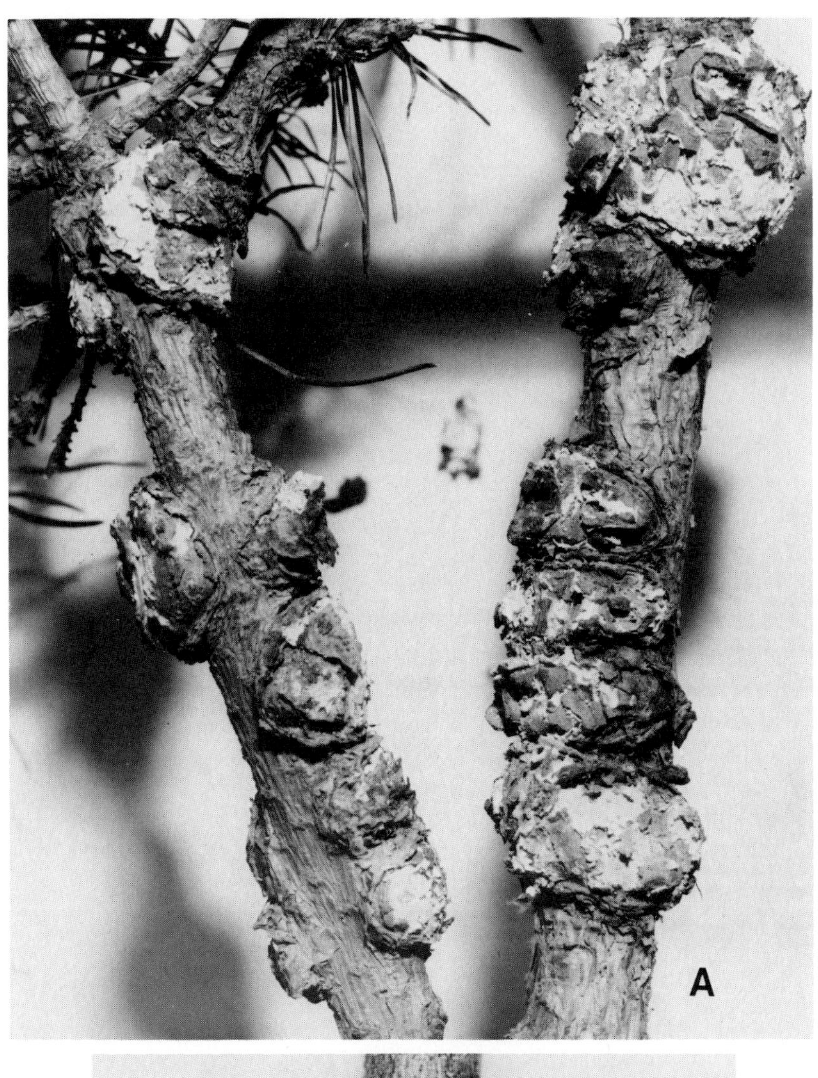

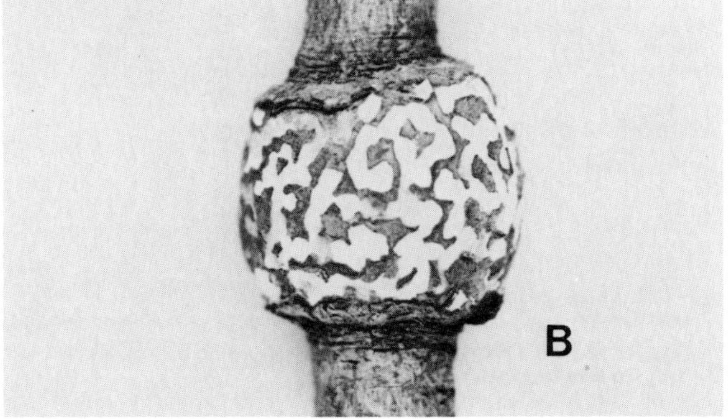

Fig. 12.11 Galls on pine stems induced by *Cronartium quercuum* covered with aecia (A, B). (Photograph A courtesy of Shade Tree Laboratories, University of Massachusetts, Amherst.)

Fig. 12.12 Telial columns of *Cronartium quercuum* on lower leaf surface of oak.

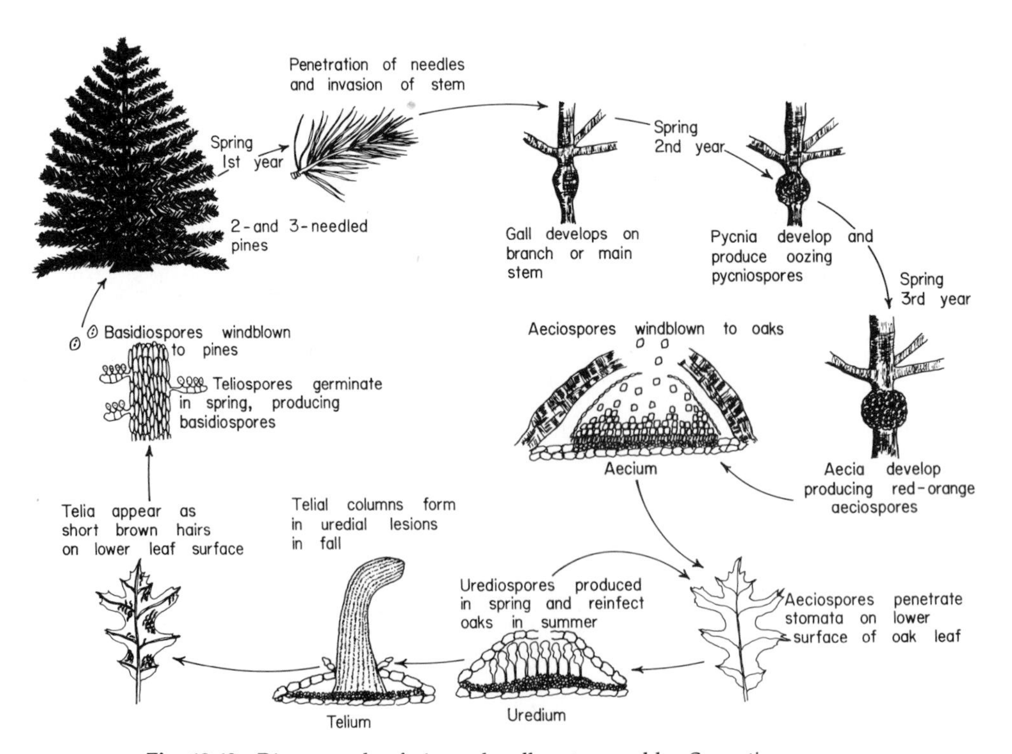

Fig. 12.13 Disease cycle of pine-oak gall rust caused by *Cronartium quercuum*.

Dwinell, L. D. (1974). Susceptibility of Southern oaks to *Cronartium fusiforme* and *Cronartium quercuum*. *Phytopathology* **64,** 400–403.

DISEASE: FUSIFORM RUST

Primary causal agent: *Cronartium fusiforme* Hedg. & Hunt

Hosts: Many 2- and 3-needled *Pinus* spp., especially loblolly (*P. taeda* L.) and slash (*P. elliottii* Engelm.) pines; *Quercus* spp. (oaks)

Symptomatology: Fusiform rust is a common problem on pines in the southern United States. Spindle-shaped galls are formed on the main stem and branches of susceptible pines (Fig. 12.14). Trees of all ages can be infected but the disease is most serious in seedlings, which are stunted and develop excessive branching around the gall (Fig. 12.15). In pole-sized or larger trees, wind

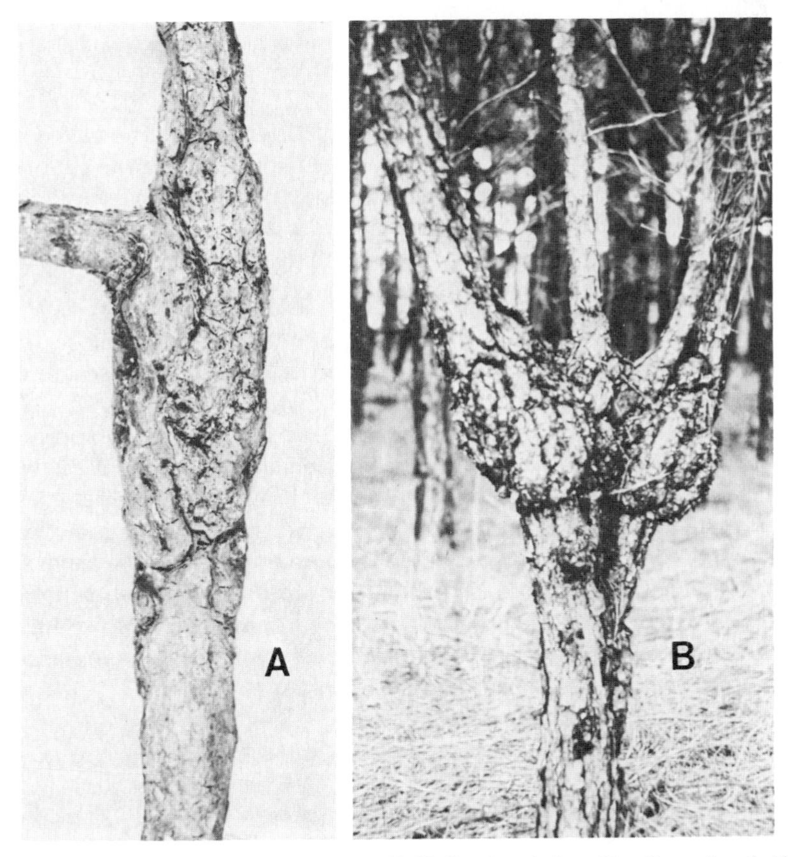

Fig. 12.14 Spindle-shaped gall on main stem of loblolly pines induced by *Cronartium fusiforme* (A, B). (Photograph B courtesy of USDA Forest Service.)

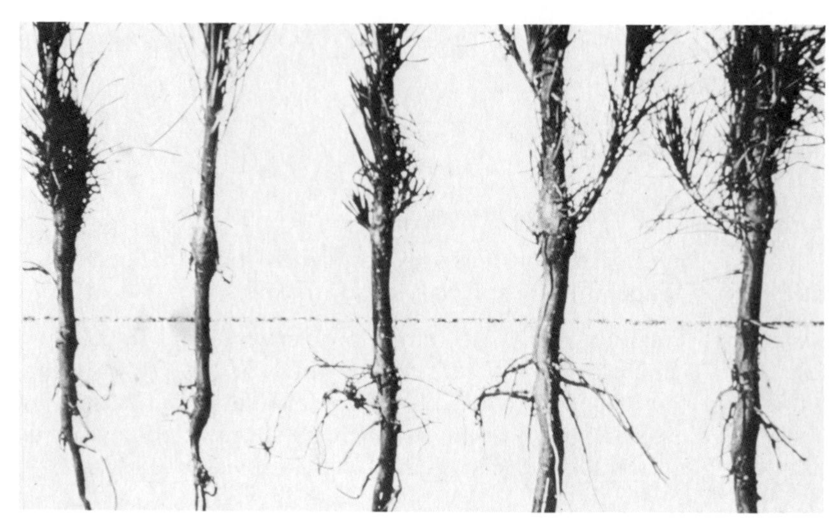

Fig. 12.15 Seedlings of loblolly pine infected with *Cronartium fusiforme*. Note spindle-shaped galls with excessive branching on lower stems. (Photograph courtesy of USDA Forest Service.)

breakage at the gall and dieback above the gall often occur. The galled regions often become sunken cankers as the trees age. In the spring the galls initially produce a yellowish ooze (pycnial exudate) and later aeciospores are produced in blisters on the gall surface. Aeciospores infect developing oak leaves and produce orange uredial lesions on the undersides. Uredia remain on the oak leaves for the rest of the season and are replaced by telial columns (Fig. 12.16) early in the next growing season.

Etiology: The pathogen is a macrocyclic heteroecious rust whose life cycle is similar to *C. quercuum* (see Fig. 12.13) and *C. ribicola* (see Fig. 12.7). Teliospores germinate in late winter or early spring and liberate basidiospores which are windblown to pines. Basidiospores enter needles and shoots directly and invade progressively to stem tissue where a perennial gall is induced. Pycnia are produced, usually beginning in the second season after infection, and spermatization occurs, followed by aecial production later in the same season. Aeciospores are windblown to expanding oak leaves where they penetrate the lower epidermis and produce uredial lesions. Urediospores are produced and can reinfect oak leaves until all leaves are fully expanded. Telia replace uredia in lesions on the oak leaves during late winter.

Control: Use the same control procedures for Fusiform rust as for pine-oak rust.

Selected References

Czabator, F. J. (1971). Fusiform rust of southern pines—A critical review. *U.S. For. Serv., Res. Pap.* **50–65.**

Jewell, F. F., True, R. P., and Mallet, S. L. (1962). Histology of *Cronartium fusiforme* in slash pine seedlings. *Phytopathology* **52,** 850–858.

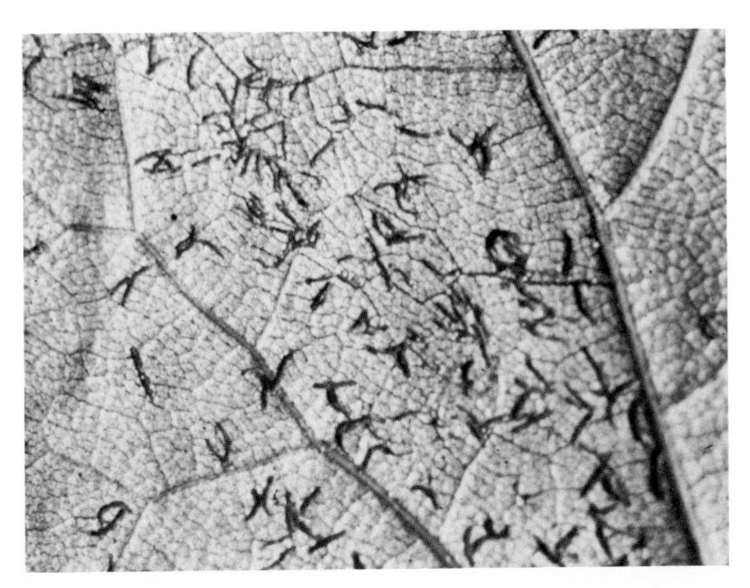

Fig. 12.16 Telial columns of *Cronartium fusiforme* on lower leaf surface of oak. (Photograph courtesy of USDA Forest Service.)

Fig. 12.17 Spots with tiny black pycnia on upper surface of apple leaf infected with *Gymnosporangium juniperi-virginianae.*

Powers, H. R., McClure, J. P., Knight, H. A., and Dutrow, G. F. (1975). Fusiform rust: Forest survey incidence data and financial impact in the South. *U.S. For. Serv., Res. Pap. SE* **SE-127.**
Snow, G. A., and Roncadori, R. W. (1965). Oak leaf age and susceptibility to *Cronartium fusiforme*. *Plant Dis. Rep.* **49,** 972–975.

DISEASE: CEDAR-APPLE RUST

Primary causal agent: *Gymnosporangium juniperi-virginianae* Schw.

Hosts: *Malus* spp. (apples and crabapples), *Juniperus* spp. (junipers, red cedar)

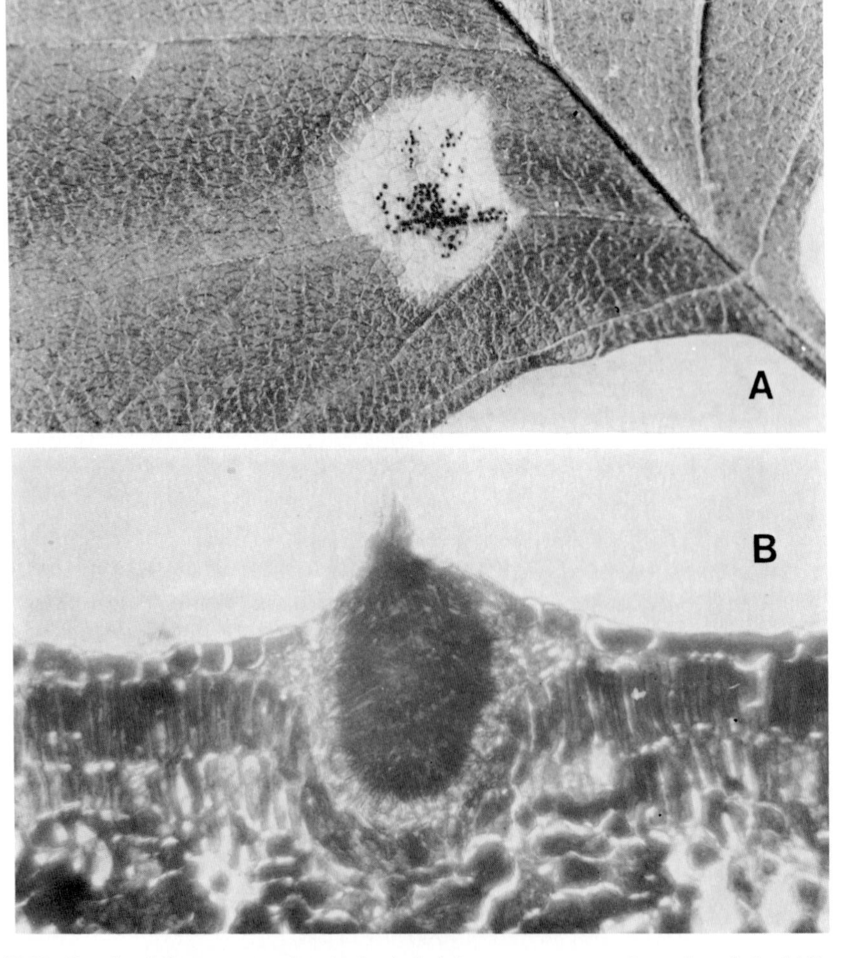

Fig. 12.18 Pycnia of *Gymnosporangium juniperi-virginianae* on upper surface of apple leaf (A), cross section of pycnium (B) 200×. (Photograph B from microscope slide prepared by Carolina Biological Supply Company, Burlington, North Carolina.)

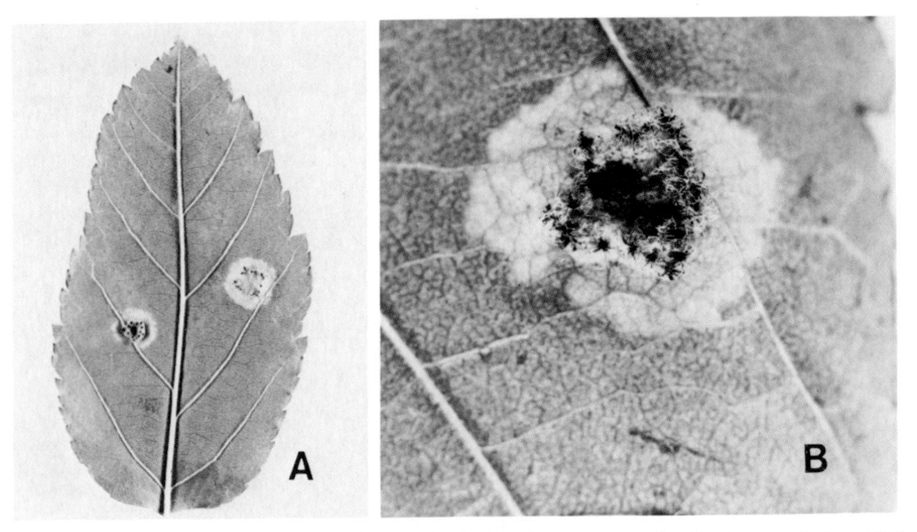

Fig. 12.19 Aecial cups of *Gymnosporangium juniperi-virginianae* surrounded by long curled peridial membranes on lower surface of apple leaf (A), closeup (B).

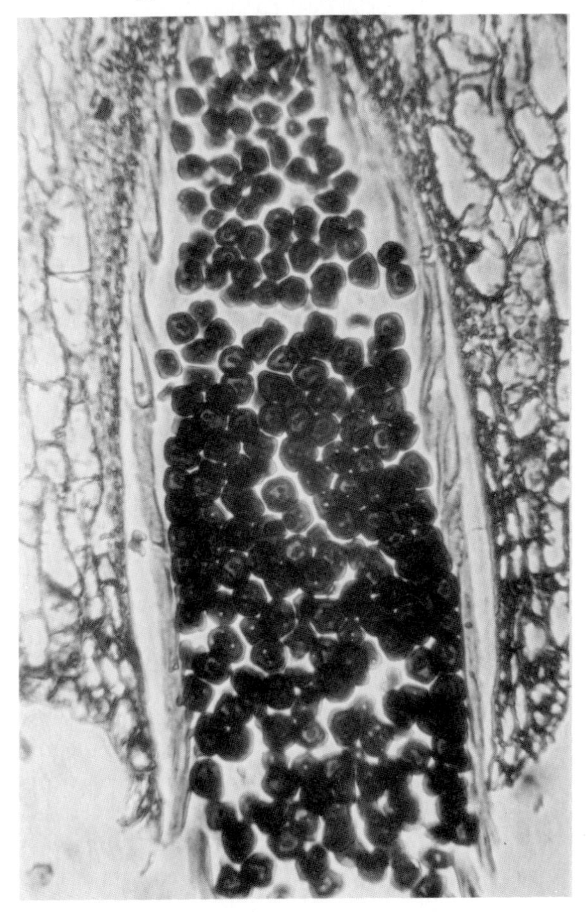

Fig. 12.20 Aeciospores in cross section of aecial cup of *Gymnosporangium juniperi-virginianae* on lower surface of apple leaf. 200×. (Photograph from microscope slide prepared by Triarch, Inc., Ripon, Wisconsin.)

Fig. 12.21 Dimpled galls induced by *Gymnosporangium juniperi-virginianae* on *Juniperus* twigs.

Symptomatology: In spring yellow spots about ½ inch (1 cm) across (Fig. 12.17) appear primarily on the upper surface of apple and crabapple leaves. The spots contain pycnia (Fig. 12.18) which, after spermatization with a compatible strain, give rise to aecia, usually on the underside of leaves. In summer, aecial cups, surrounded by long curled peridial membranes (Fig. 12.19) liberate yellow aeciospores (Fig. 12.20). On the growing *Juniperus* twigs brown dimpled galls of varied size up to two inches (5 cm) across are produced (Fig. 12.21). During the early spring of the second year of development, about the time of bud break for apples and crabapples, the dimples begin to expand to form small spikes (Fig. 12.22). During wet periods for approximately the next four to six weeks these spikes, which are developing telia, greatly expand to form a mass of soft yellow-orange tendrils, often several inches long (Fig. 12.23).

Etiology: The pathogen is a heteroecious demicyclic rust (lacks urediospores) (Fig. 12.24). Basidiospores from germinating telia infect young *Malus* spp. foliage by direct penetration. Pycnia, which ooze pycniospores, are formed and then spermatization with a compatible strain occurs. Aecia are then formed and aeciospores are windblown to foliage of *Juniperus* spp. where they penetrate young shoots. The fungus remains in a vegetative state for the next

Fig. 12.22 Small telial spikes on galls induced by *Gymnosporangium juniperi-virginianae* on *Juniperus* twigs. (Photograph courtesy of Shade Tree Laboratories, University of Massachusetts, Amherst.)

20 months while a gall develops on the juniper twig. Telia form on the outside of the gall (Fig. 12.25) during the early spring of the second growing season after infection and mature teliospores release basidiospores during wet periods. Cedar-hawthorn rust and cedar-quince rust are similar rust diseases that occur on *Juniperus* spp. and on hawthorn and quince, respectively. The symptomatology and etiology of both these diseases are similar to cedar-apple rust.

Control: Protect foliage of apples and crabapples with fungicide applications during leaf development. *Juniperus* spp. can also be protected by fungicide

Fig. 12.23 Mature telial tendrils of *Gymnosporangium juniperi-virginianae* on *Juniperus* twigs. (Photographs courtesy of David Gadoury, University of New Hampshire, Durham.)

applications in midsummer until late summer. Remove one host, either *Malus* spp. or *Juniperus* spp., from areas where this disease is a problem. Avoid planting both genera in the same local area. Select resistant varieties of *Malus* or *Juniperus*.

Selected References

Bliss, D. E. (1933). The pathogenicity and seasonal development of *Gymnosporangium* in Iowa. *Iowa Agr. Exp. Sta. Bull.* 166.

Cromwell, I. H. (1934). The hosts, life history, and control of the cedar-apple rust fungus *Gymnosporangium juniperi-virginianae*. *J. Arnold Arbor., Harvard Univ.* **15,** 163–232.

Palmiter, D. H. (1952). Rust diseases of apples and their control in the Hudson Valley. *N.Y. Agric. Exp. Stn., Bull.* No. 756.

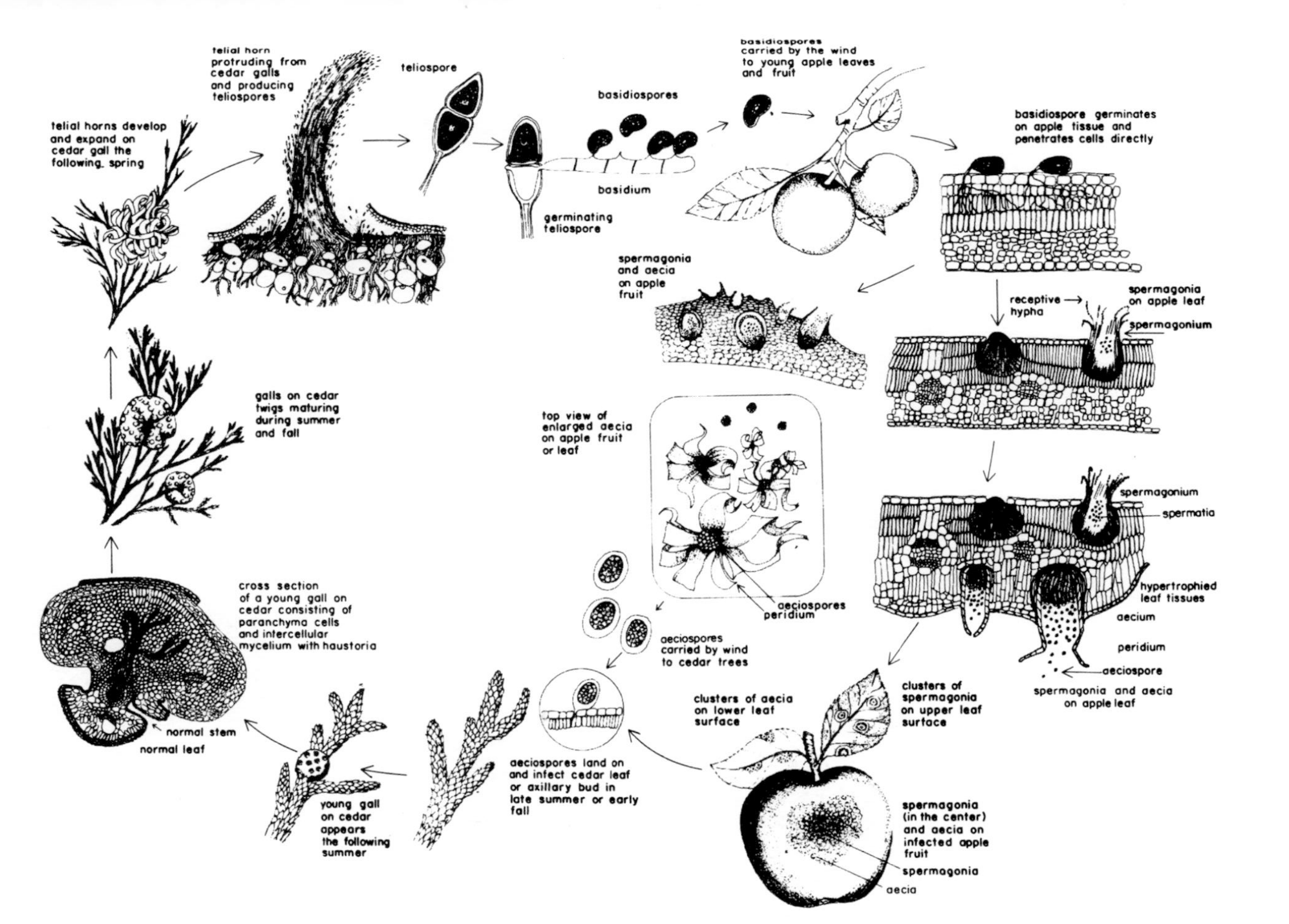

Fig. 12.24 Disease cycle of cedar-apple rust caused by *Gymnosporangium juniperi-virginianae*. (Drawing courtesy of George N. Agrios, "Plant Pathology," 2nd ed. Academic Press, New York, 1978.)

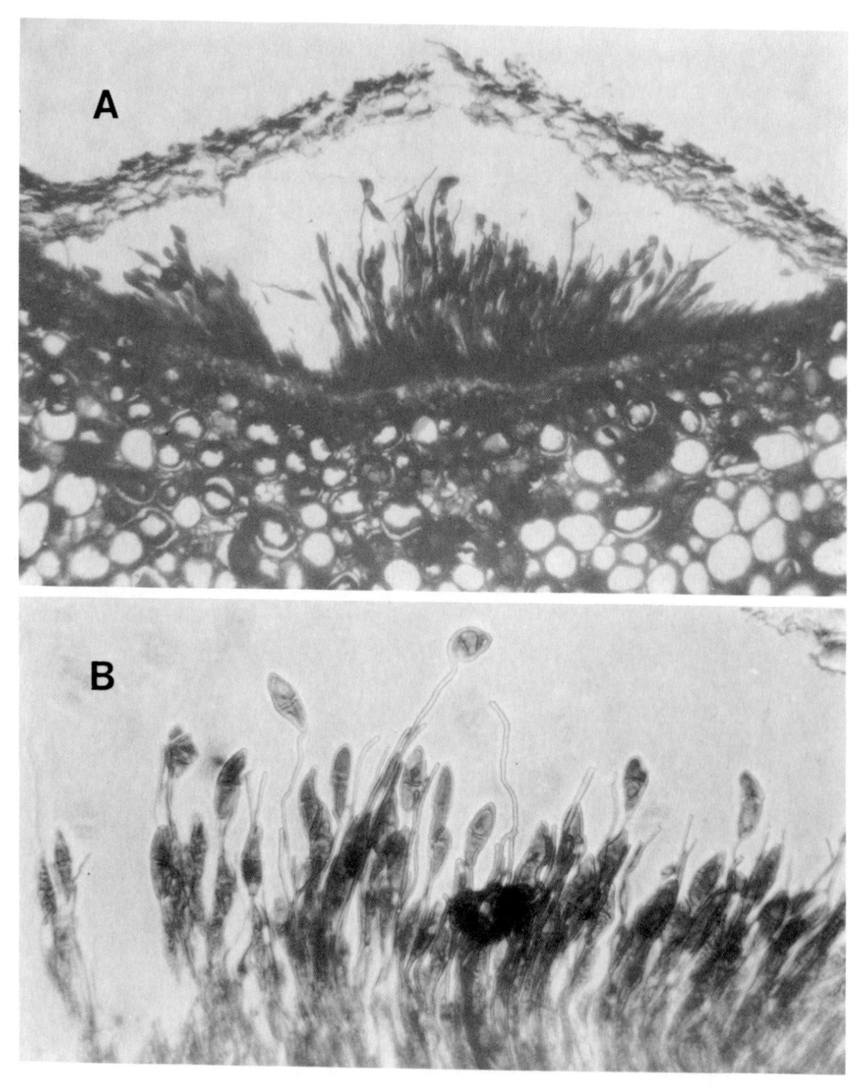

Fig. 12.25 Cross section of telium (A) and teliospores (B) of *Gymnosporangium juniperi-virginianae*. A, 80×; B, 200×. (Photographs from microscope slide prepared by Carolina Biological Supply Company, Burlington, North Carolina.)

DISEASE: ASH RUST

Primary causal agent: Puccinia peridermiospora (Ell. & Tr.) Arth.

Hosts: Ash (*Fraxinus* spp.), marsh grass (*Spartina* spp.)

Fig. 12.26 Pycnia on upper surface of ash leaflet infected with *Puccinia peridermiospora*. (Photograph courtesy of Shade Tree Laboratories, University of Massachusetts, Amherst.)

Symptomatology: Yellow-orange pustules (pycnia) form on ash leaflets (Fig. 12.26), petioles, and sometimes shoots and seeds. These are followed by orange cluster cups (aecia) on bottoms of leaves and on stems around pycnia (Fig. 12.27). Defoliation occurs in summer following heavy infection (Fig. 12.28); refoliation may then occur in late summer. In summer the leaf epidermis of marsh grasses is ruptured to form red-orange lesions (Fig. 12.29) which contain urediospores. Later, lesions become darker, appearing red-black, and teliospores (Fig. 12.30) replace urediospores during the fall.

Etiology: Basidiospores from infected marsh grass are windblown to developing ash leaves in early spring (Fig. 12.31). Following penetration of the ash leaf, pycnia are produced around the site of infection. Spermatization between compatible strains is accomplished with the aid of insects which are attracted

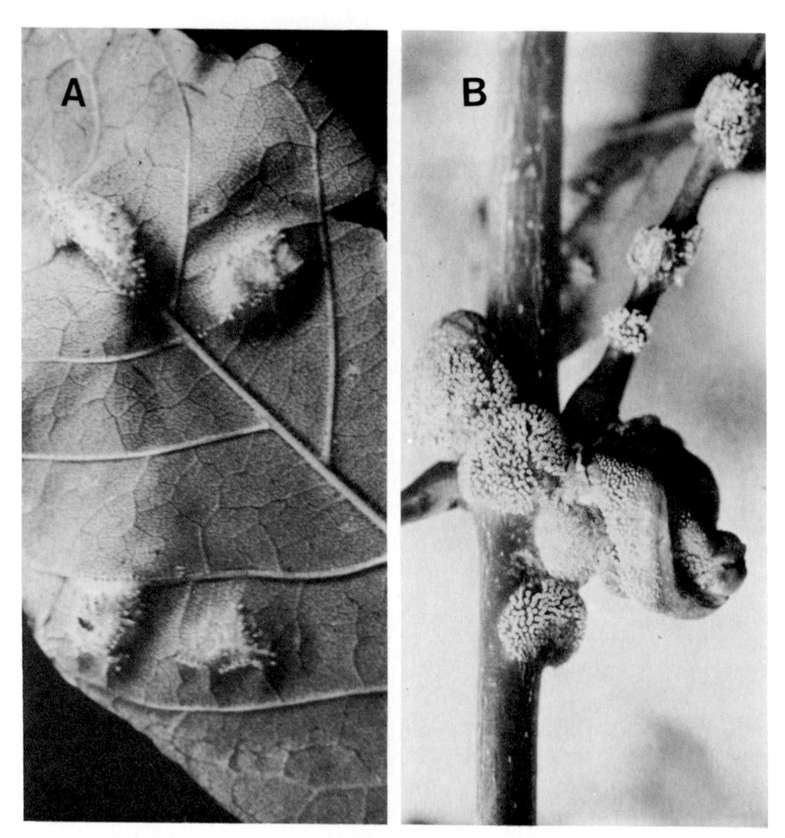

Fig. 12.27 Aecia on lower surface of ash leaflet (A) and ash petioles and twig (B) infected with *Puccinia peridermiospora*.

to the pycnial ooze, and a dikaryotic mycelium is formed. Aecia are produced on the lower leaf surface or adjacent to the pycnia, and binucleate aeciospores are produced. Aeciospores are windblown to the marsh grass and enter the leaves via the stomata. After successful infection, a lesion is formed and single-celled urediospores are produced. These spores can reinfect marsh grass throughout the summer. By late summer or early fall two-celled teliospores have replaced urediospores in the lesions. In *S. alterniflora,* at least in some regions, urediospores may persist over winter and may be able to infect marsh grass in the spring. The teliospores remain dormant during winter and germinate to produce basidiospores during wet periods in the spring.

Control: Protectant fungicide applications have been successful, if applied as leaves emerge and develop in the spring. However, the size of mature ash trees may make this an expensive operation and useful only on selected shade trees. Annual control is not necessary, since the disease only becomes severe

Fig. 12.28 White ash trees defoliated by ash rust.

Fig. 12.29 Elongate uredia on leaf of marsh grass infected with *Puccinia peridermiospora* (A), closeup (B).

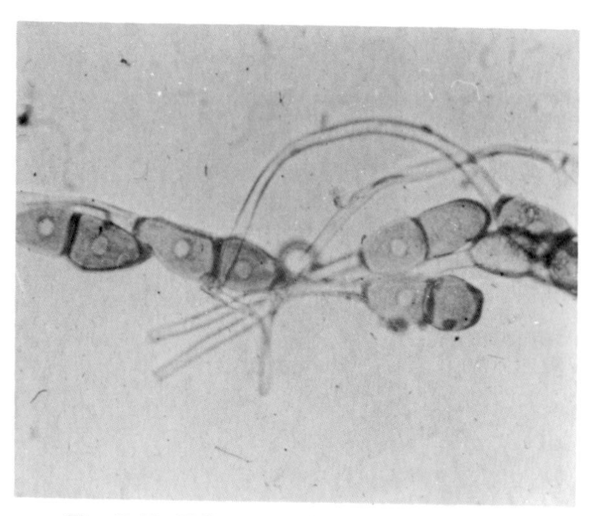

Fig. 12.30 Teliospores of *Puccinia peridermiospora*.

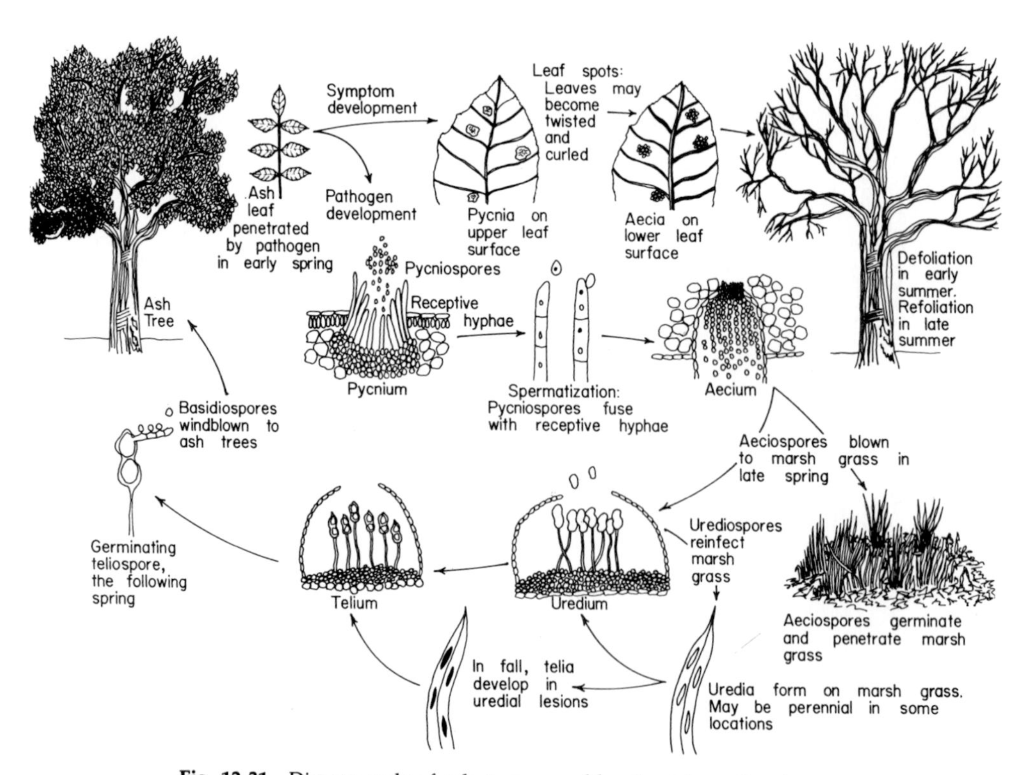

Fig. 12.31 Disease cycle of ash rust caused by *Puccinia peridermiospora*.

Fig. 12.32 Salt marsh grasses on edge of estuary.

every 15–20 years. Planting of ash, particularly white ash, should be avoided near the seacoast around salt marshes (Fig. 12.32).

Selected References

Blanchard, R. O. (1974). Ash leaf rust. *For. Notes* **118,** 27–28.

Partridge, A. D. (1957). The ash leaf rust syndrome in New Hampshire: Suscepts, incitant, epidemiology and control. Ph.D. Thesis, Univ. of New Hampshire, Durham.

13

Root Diseases

INTRODUCTION

Diseases of the root system are some of the most serious problems of forest, shade, and nursery trees. The major root diseases of trees are caused by soilborne fungi that are capable of persisting long periods as saprophytes on dead plant materials. The ability of root disease fungi to penetrate roots and colonize the root system is often closely related to general health of the tree. In general, trees exhibiting good growth and vigor are resistant to root diseases while those suffering from stress or disturbances and/or in poor vigor are most susceptible to root disease. Annosus root rot is one exception.

The aboveground symptoms of most root diseases are similar, such as a general decline in crown condition, poor growth rate, and poor condition of foliage. Belowground symptoms may vary somewhat depending on the infecting fungus. *Phytophthora* spp. are phycomycetous fungi which cause root lesions and eventually kill the roots. *Heterobasidion annosum* and *Armillaria mellea* are both basidiomycetous fungi which kill the roots and then proceed to decay them.

Selected References

Epstein, A. H. (1978). Root graft transmission of tree pathogens. *Annu. Rev. Phytopathol.* **16,** 181–192.

Garrett, S. D. (1970). "Pathogenic Root-Infecting Fungi." Cambridge Univ. Press, London and New York.

DISEASE: ARMILLARIA ROOT ROT (SHOESTRING ROOT ROT)

Primary causal agent: Armillaria mellea (Vahl. ex Fr.) Kummer

Hosts: Most woody plants are susceptible

Symptomatology: Affected trees usually exhibit decreased crown growth, dieback, and general decline as initial responses to the disease (Fig. 13.1). Sometimes, however, an apparently healthy tree may die in a few weeks. The pathogen causes death and decay of roots, but also causes cankers around the root collar which often girdle the tree and cause its death. Abundant resin flow is common. Infected trees are also subject to windthrow due to weakened support of the root system. Trees with root injury, recently transplanted trees, or trees that have suffered severe stress, such as prolonged drought or insect defoliation, are often attacked by *A. mellea.* Under the bark of infected trees a necrotic zone or killing front can be seen at the junction of healthy and dis-

Fig. 13.1 Declining maple tree infected with *Armillaria mellea.* (Photograph courtesy of Philip M. Wargo, USDA Forest Service, Hamden, Connecticut.)

eased tissues. Behind this front white mycelial fans are often formed on the xylem surface (Fig. 13.2). Dark brown to black strands of fungus mycelium, termed rhizomorphs or shoestrings, often form under the bark or around the outside of the root (Fig. 13.3). In the early fall honey-colored mushrooms may appear at the base of an infected tree or around trees recently killed by the disease (Fig. 13.4). Although the mushroom stage is short-lived, the shriveled mushrooms can often be detected for several months.

Etiology: The pathogen is a persistent soil fungus that is common to most forest soils worldwide (Fig. 13.5). It can persist as a saprophyte in a vegetative state indefinitely in dead roots and can migrate to roots of healthy trees via rhizomorphs. These structures are thought to function in transport of materials to the growing regions of the fungus. After penetration of the healthy

Fig. 13.2 White mycelial fan of *Armillaria mellea* on xylem surface of pine. (Photograph courtesy of Philip M. Wargo, USDA Forest Service, Hamden, Connecticut.)

Fig. 13.3 Rhizomorphs of *Armillaria mellea.*

tree, the pathogen progressively invades the roots and buttress area where it kills the cambium and later decays the xylem beneath. The ability of the fungus to penetrate and progressively invade roots is usually directly related to the health of the tree. During early fall the sexual basidiocarp (mushroom) stage develops and liberates windblown basidiospores, which can become established on recent root or buttress wounds or on dead woody tissue.

Control: Prevention can best be achieved by helping trees avoid stresses such as moisture imbalance, insect defoliation, and people pressures, that predispose trees to infection. When planting trees, select planting sites that allow for vigorous root growth. Where possible, avoid competition from lawn grasses.

Selected References

Leaphart, C. D. (1963). *Armillaria* root rot. *U.S. For. Serv., For. Pest Leafl.* No. 78.

Pawsey, R. G., and Rahman, M. A. (1976). Chemical control of infection by honey fungus, *Armillaria mellae:* A review. *J. Arboricult.* **2,** 161–169.

Fig. 13.4 Fruiting bodies of *Armillaria mellea,* mature (A), young (B). (Photographs courtesy of Philip M. Wargo, USDA Forest Service, Hamden, Connecticut.)

Raabe, R. D. (1962). Host list of the root-rot fungus, *Armillaria mellea*. *Hilgardia* **33,** 25.

Shaw, C. G., III, Lewis, F., Rolph, L., and Hunt, J. (1976). Dynamics of pine and pathogen as they relate to damage in a forest attacked by *Armillaria*. *Plant Dis. Rep.* **60,** 214–218.

Thomas, H. E. (1934). Studies on *Armillaria mellea* (Vahl.) Quel., infection, parasitism, and host resistance. *J. Agric. Res.* **48,** 187–218.

Wargo, P. M. (1972). Defoliation-induced chemical changes in sugar maple roots stimulate growth of *Armillaria mellea*. *Phytopathology* **62,** 1278–1283.

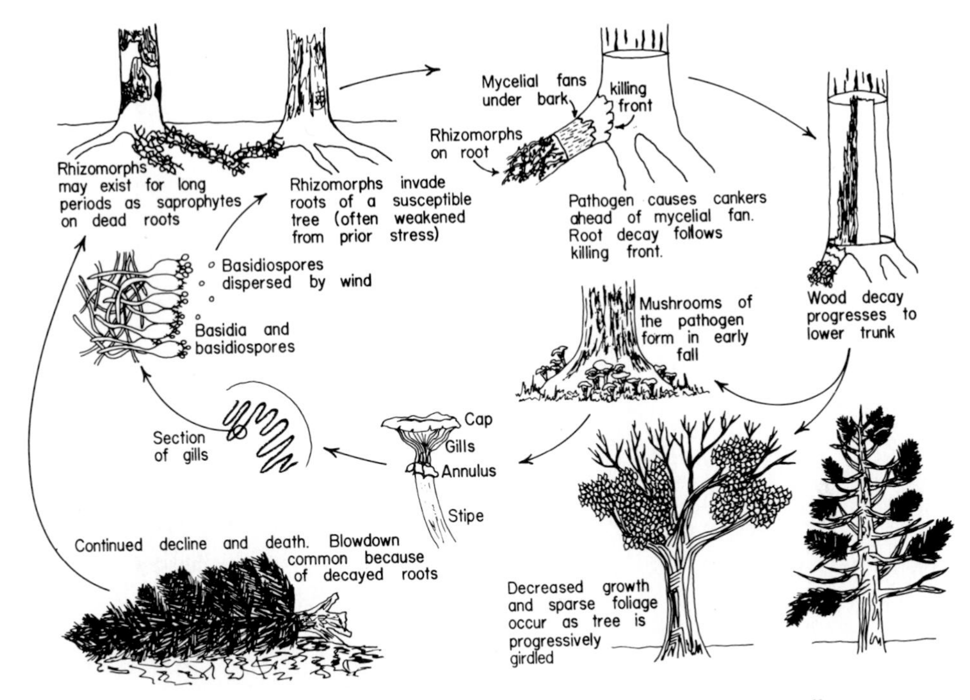

Fig. 13.5 Disease cycle of Armillaria root rot caused by *Armillaria mellea*.

DISEASE: ANNOSUS ROOT ROT

Primary causal agent: *Heterobasidion annosum* (Fr.) Bref. (formerly *Fomes annosus* (Fr.) Cke.)

Hosts: All conifers are susceptible

Symptomatology: Affected trees exhibit poor growth, produce short needles, and hold them for only the current season. Abundant cone crops may also be seen. As the disease progresses, the foliage continues to thin, tip burn may be evident and eventually the tree turns completely brown and dies. Sometimes infected trees will blow over while still alive (Fig. 13.6). Extensive decay in the roots and buttress areas can usually be found in the blown down trees and often in the dead standing trees as well. Dead trees may also exhibit excessive resin soaking on the trunk. The disease often occurs in circular areas and radiates outward killing trees on the periphery. The basidiocarp of *H. annosum* can usually be found in the duff layer at the base of dead trees and in trees in the advanced stages of the disease (Fig. 13.7). They have a dark brown top and

Fig. 13.6 Loblolly pines infected with *Heterobasidion annosum* showing sparse foliage and windthrow of living trees. (Photograph courtesy of W. A. Stegall, C. S. Hodges, E. Ross, and E. G. Kuhlman, USDA Forest Service, Research Triangle Park, North Carolina.)

a white pore layer underneath. Sporophores also form occasionally on roots and stumps and on coniferous slash.

Etiology: Basidiospores are windblown to freshly cut stumps of susceptible conifers (Fig. 13.8). The pathogen grows through the stumps into the root system where it can invade healthy roots from adjacent trees via root grafts. Healthy roots are killed and decayed as the pathogen advances into the buttress area. The infected tree may be blown down while still alive due to instability caused by decayed roots, or the tree may be killed from trunk girdling by the pathogen and remain standing. *Heterobasidion annosum* will continue to invade and kill adjacent trees in a roughly circular pattern for several years. Sporophores are usually produced at the base, on roots, and stumps of infected trees and sometimes on coniferous slash. Basidiospores are released through pores at the base of the basidiocarp and serve as inoculum to establish new infection

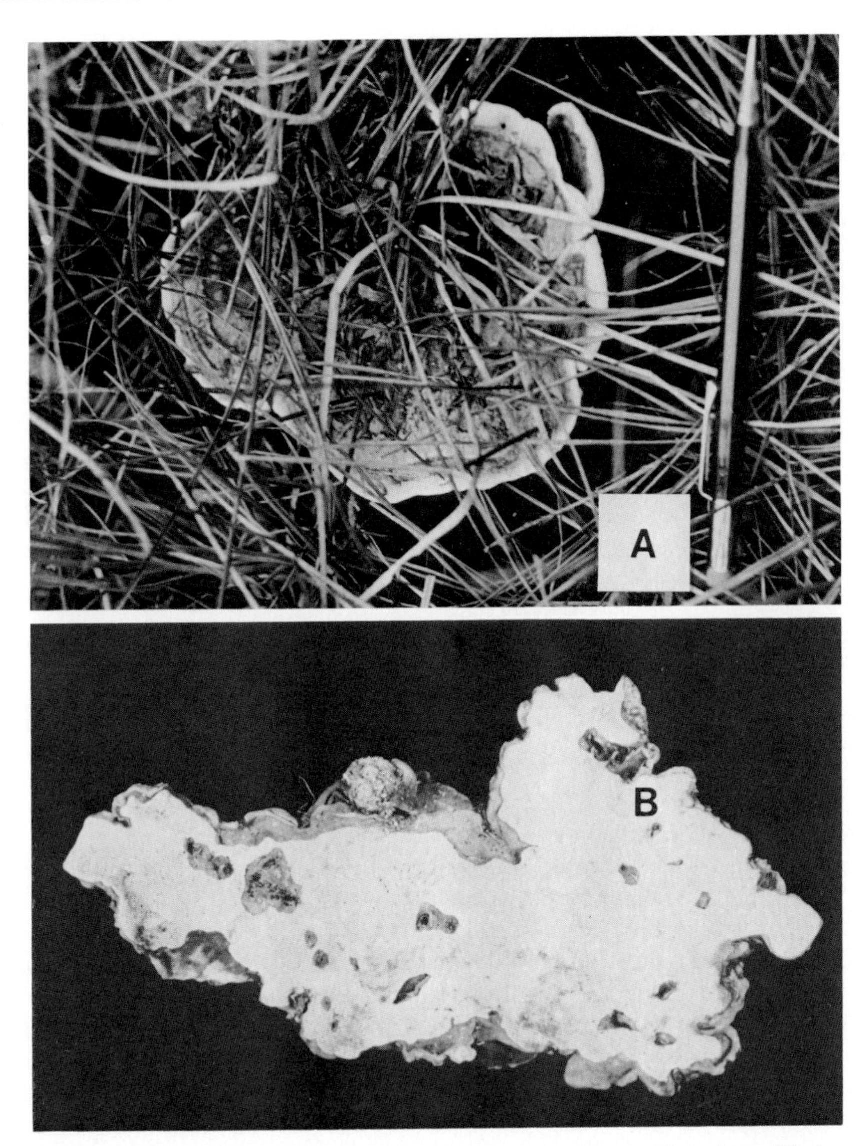

Fig. 13.7 Basidiocarp of *Heterobasidion annosum*, upper (A) and lower (B) surfaces. (Photograph A courtesy of W. A. Stegall, C. S. Hodges, E. Ross, and E. G. Kuhlman, USDA Forest Service, Research Triangle Park, North Carolina.)

centers. The pathogen also produces an asexual state (*Oedocephalum*) whose role in the disease cycle is unclear, but is useful in laboratory identification of this fungus in culture.

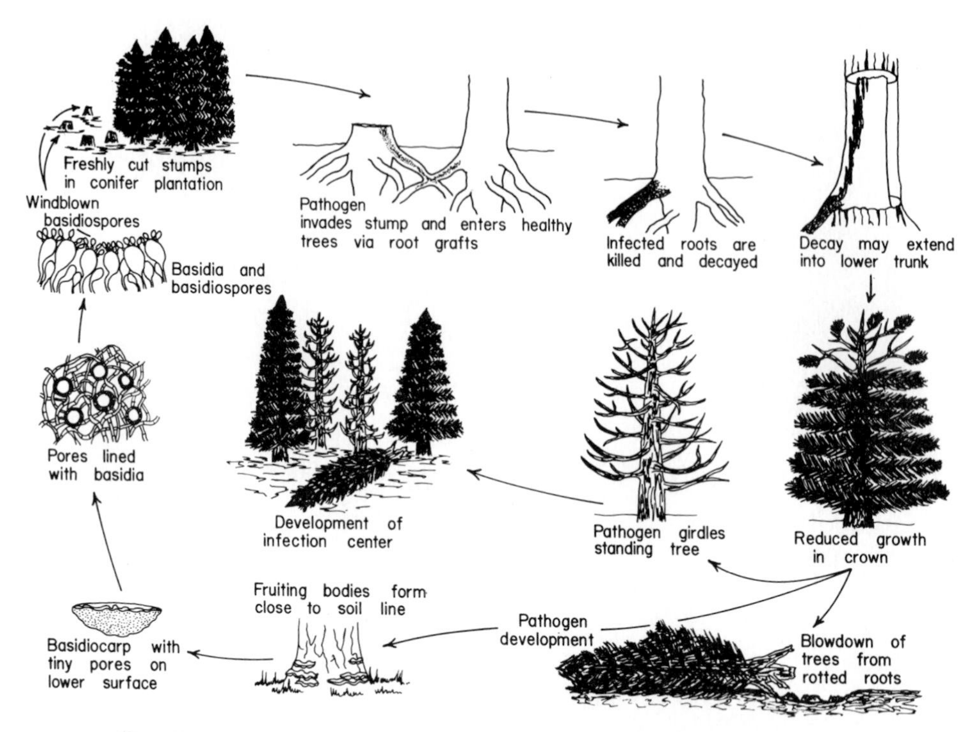

Fig. 13.8 Disease cycle of Annosus root rot caused by *Heterobasidion annosum.*

Control: Assess hazard potential:

High hazard—well drained soil with medium to high fertility, often former agricultural land.

Low hazard—poorly drained soils with low fertility.

On high hazard sites, protect freshly cut stumps of conifers with application of borax, urea, or the wood rotting fungus *Peniophora gigantea.* Reduce the number of thinnings per rotation. Thin in summer in the southern United States (below 34° N latitude).

On low hazard sites, preventative treatments usually are not warranted.

Selected References

Hodges, C. S. (1974). Symptomatology and spread of *Fomes annosus* in southern pine plantations. *U.S. For. Serv., Res. Pap. SE* **SE-114.**

Kuhlman, E. G., ed. (1973). *Procs. Int. Congr.* Fomes annosus, *4th, IUFRO, USDA For. Serv., Washington, D.C.*

Kuhlman, E. G., Hodges, C. S., and Froelich, R. C. (1976). Minimizing losses to *Fomes annosus* in the southern United States. *U.S. For. Serv., Res. Pap. SE* **SE-151.**

Smith, R. S. (1970). Borax to control *Fomes annosus* infection of white fir stumps. *Plant Dis. Rep.* **54,** 872–875.

Fig. 13.9 Sweet cherry trees infected with Phytophthora root and crown rot. Seven-year-old cultivar Royal Ann trees on Mahaleb rootstock in a heavily infected commercial orchard (A). Failure to leaf out in spring (B). Collapse (wilt) in early summer (C). Necrosis under bark in lower stem (D). Trunk canker (E). Note sunken bark areas on lower trunk (d) and upper margin of canker (m). (Photographs courtesy of S. M. John Mircetich, USDA-SEA, University of California, Davis.) [From S. M. Mircetich and M. E. Matheron, *Phytopathology* **66,** 549–558 (1976).]

DISEASE: PHYTOPHTHORA ROOT ROTS

Primary causal agents: *Phytophthora cinnamomi* Rands and *P. cactorum* (Leb. & Cohn) Schr. are most common, but other species of *Phytophthora* may be involved.

Hosts: Most woody plants of all ages are susceptible

Symptomatology: Affected trees exhibit poor growth and display general symptoms of decline, such as sparse, tufted foliage which is light green to yellow, branch dieback, and substantial reduction in incremental growth rate (Fig. 13.9, 13.10). Necrosis of small roots occurs initially and is followed by the formation of brown to black lesions on larger roots (Fig. 13.11). Disease progression may result in decline over several seasons or killing within one or two seasons. This disease is often more severe in poorly drained areas and in soils of low fertility, or in areas that have been disturbed, such as by house or road construction. On some hosts *Phytophthora* spp. also produce diffuse cankers on the trunk and branches. Infected seedlings in the nursery may show no

Fig. 13.10 Sand pine infected with *Phytophthora cinnamomi* (left) and healthy sand pine (right). (Photograph courtesy of W. A. Stegall, C. S. Hodges, E. Ross, and E. G. Kuhlman, USDA Forest Service, Research Triangle Park, North Carolina.)

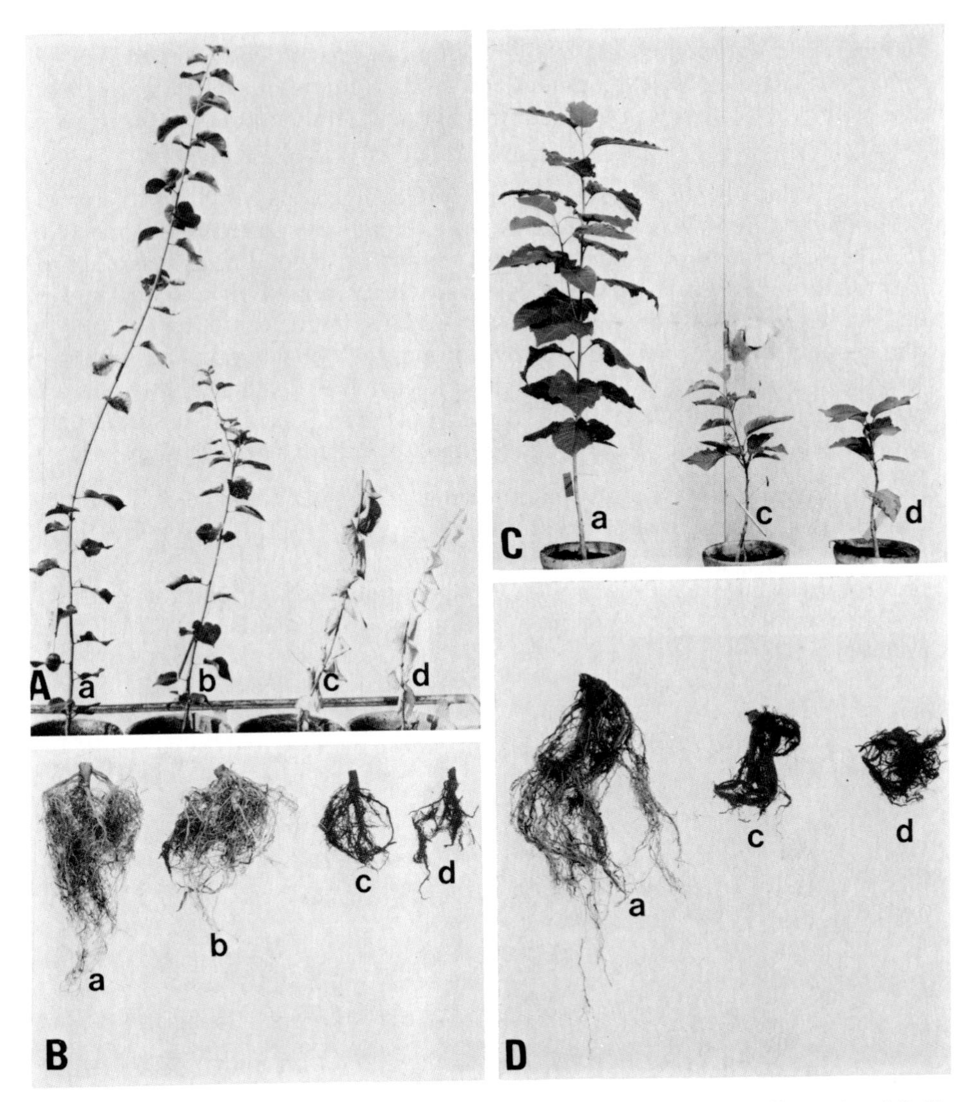

Fig. 13.11 Shoots and roots of healthy and *Phytophthora*-infected cherry seedlings. A and B. Six-month-old seedlings, shoots (A) and roots (B), noninfested soil (a) and soil infested with *Phytophthora drechsleri* (b), *P. megasperma* (c), and *P. cambivora* (d). C and D. Similar treatments with one-year-old seedlings except treatment (b) not included. Note the drastic reduction in shoot and root growth due to infection by *Phytophthora* spp. (Photographs courtesy of S. M. John Mircetich, USDA-SEA, University of California, Davis. [From S. M. Mircetich and M. E. Matheron, *Phytopathology* **66,** 549–558 (1976).]

aboveground symptoms due to their optimum growth environment but exhibit poor lateral root development and contain numerous dead roots. However, survival rate of diseased seedlings following transport out of the nursery is low.

Etiology: *Phytophthora* spp. are soil-borne phycomycetous fungi which are able to survive indefinitely in dead host tissue as chlamydospores or oospores (Fig. 13.12). During favorable moisture and temperature conditions these spores germinate and liberate motile zoospores which can swim, with the aid of flagella, a short distance to roots of a tree. The fungus can enter the tree by direct penetration of a young root or though a root wound. The pathogen progressively invades the root system until the tree declines and eventually dies. The fungus produces thick-walled oospores (sexual state) and thick-walled chlamydospores during conditions adverse for vegetative growth.

Control: Promote tree vigor by maintaining good soil drainage and fertility. Check roots of seedlings to ensure the transplant of only disease-free stock from the nursery. Avoid replanting after a tree has died from a Phytophthora root rot unless the dead root system has been removed and the soil around it has been fumigated. Fumigate any nursery bed where diseased stock has been found.

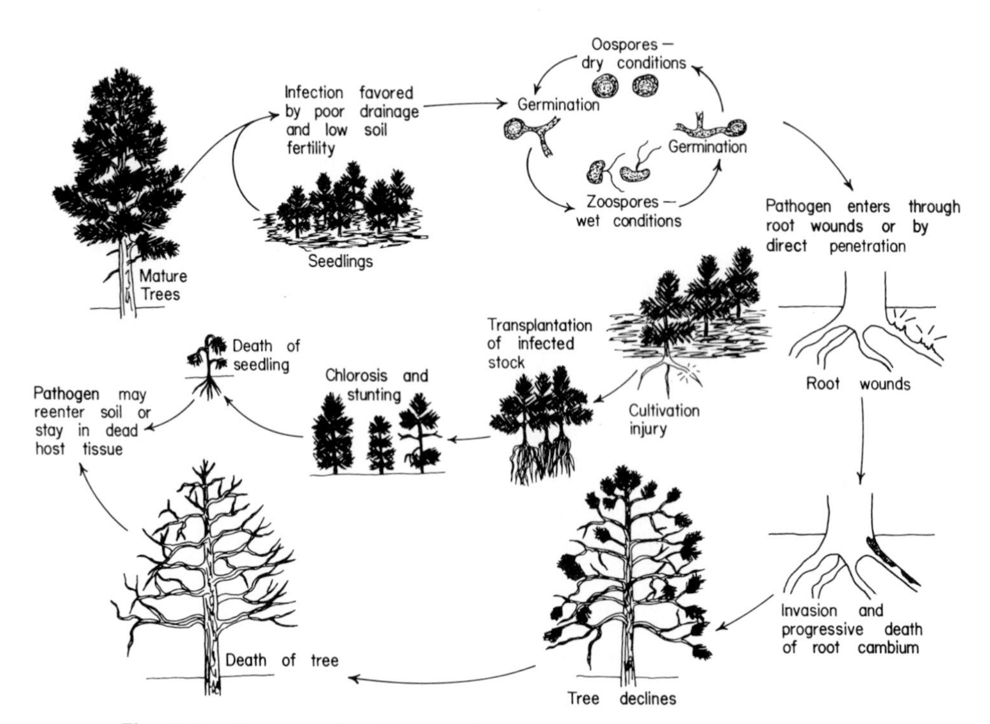

Fig. 13.12 Disease cycle of Phytophthora root rot caused by *Phytophthora* spp.

Selected References

Campbell, W. A., and Copeland, O. L. (1954). Littleleaf disease of shortleaf and loblolly pines. *U.S. Dep. Agric., Circ.* No. 940.

Marx, D. H. (1972). Ectomycorrhizae as biological deterrants to pathogenic root infections. *Annu. Rev. Phytopathol.* **10,** 429–454.

Marx, D. H., and Davey, C. B. (1969). The influence of ectotrophic mycorrhizal fungi on the resistance of pine roots to pathogenic infection IV. Resistance of naturally occurring mycorrhizae to infections by *Phytophthora cinnamomi. Phytopathology* **59,** 559–565.

Roth, L. E., Bynum, H. H., and Nelson, E. E. (1972). Phytophthora root rot of Port-Orford-cedar. *U.S. For. Serv., For. Pest Leafl.* No. 131.

Roth, L. F., and Kuhlman, E. G. (1966). *Phytophthora cinnamomi* an inhibiting threat to Douglas-fir forestry. *For. Sci.* **12,** 147–163.

Zentmyer, G. A. (1980). "*Phytophthora cinnamomi* and the Diseases It Causes," Monograph No. 10. Am. Phytopath. Soc., St. Paul, Minnesota.

MYCORRHIZAE

Mycorrhizae are fungus-feeder root associations in which the fungus is actually invasive in host cells and/or tissues. From this standpoint, the fungus could be considered pathogenic. However, mycorrhizal roots are considered to play a

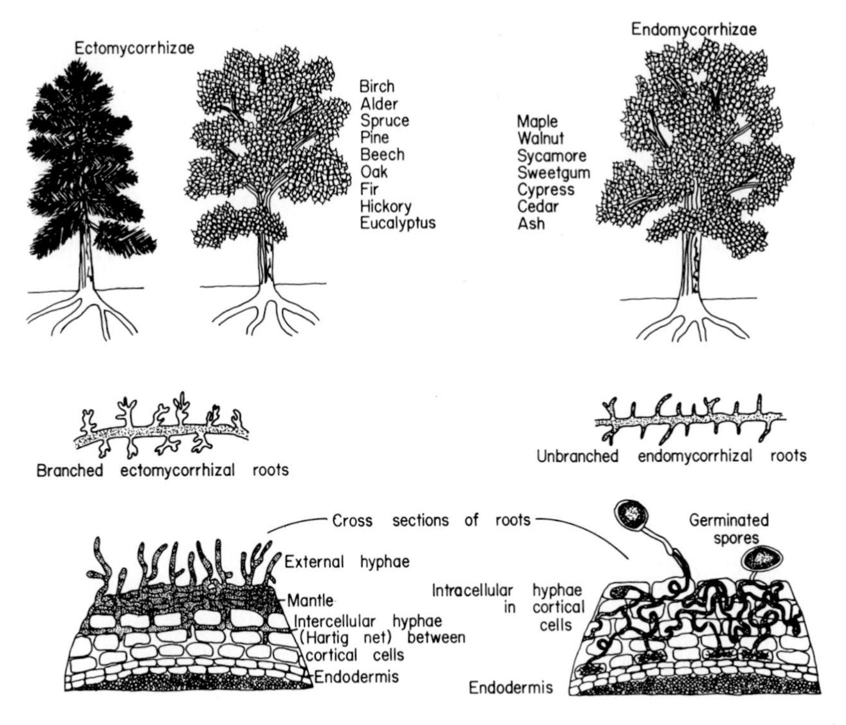

Fig. 13.13 Mycorrhizae on tree roots.

Fig. 13.14 Ectomycorrhizae on pine roots. (Photograph courtesy of Botany and Plant Pathology Dept., University of New Hampshire, Durham.)

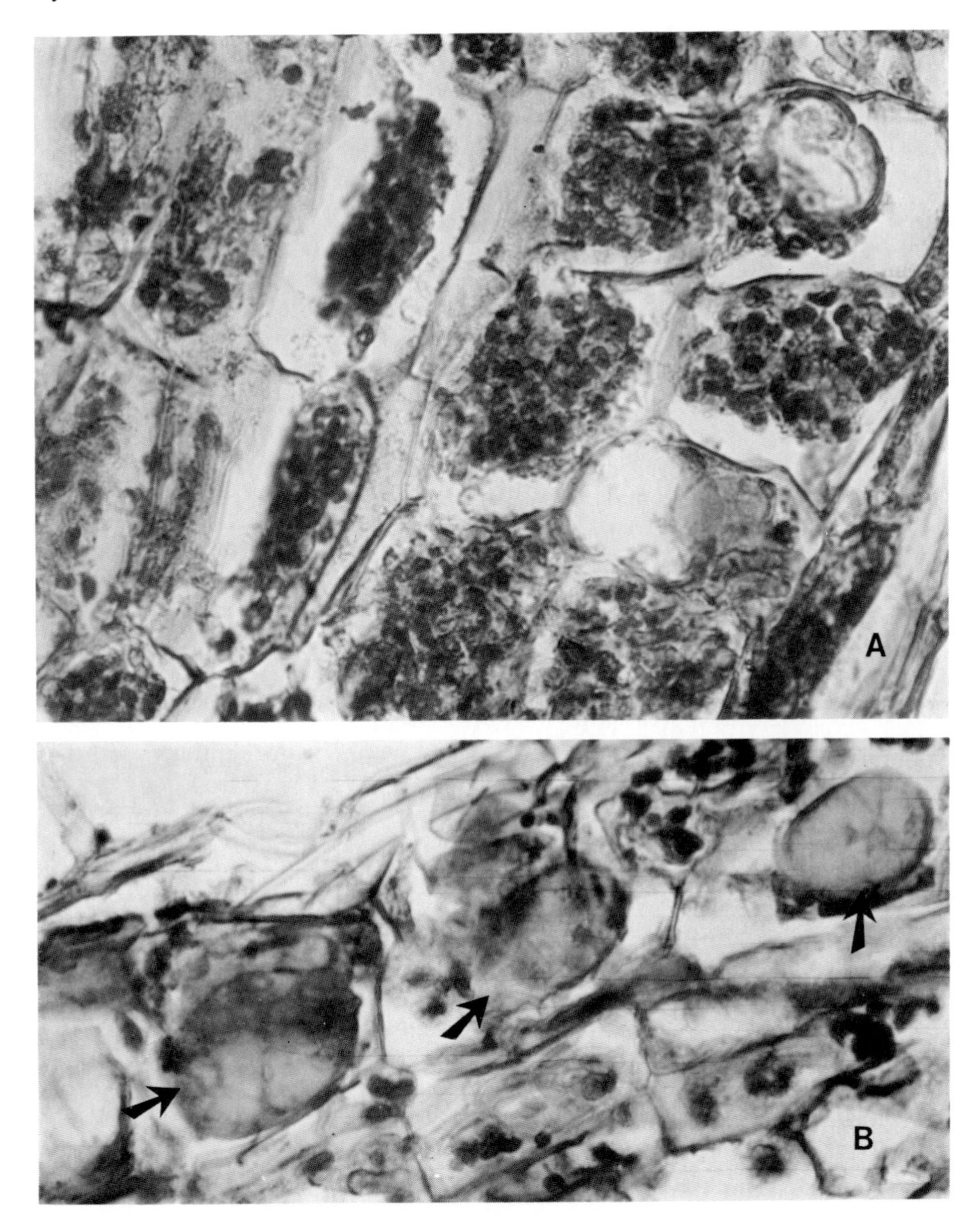

Fig. 13.15 VA mycorrhizae roots of sugar maple showing arbuscules (A) and vesicles (B) in root cortex cells. (Photographs courtesy of Terry A. Tattar, University of Massachusetts, Amherst.) [From R. A. Spitko, T. A. Tattar, and R. A. Rohde, *Can. J. For. Res.* **8,** 375–379 (1978).]

major role in root absorption of soil nutrients and water. Early establishment of abundant mycorrhizal associations in transplanted trees is critical for rapid development and continued growth, and in many cases for protection against pathogenic soil fungi. Mycorrhizae are now known to be ubiquitous in the roots of trees and also in the roots of almost all higher plants.

There are two major groups of mycorrhizae: (1) ectomycorrhizae and (2) vesicular-arbuscular mycorrhizae (endomycorrhizae) (Fig. 13.13). Ectomycorrhizae are the most easily seen due to their modification of the anatomy of the feeder roots (Fig. 13.14). Ectomycorrhizal roots have a fungus covering (mantle) over the outside of the feeder root and intercellular hyphae throughout the cortex (Hartig net). These structures are thought to present barriers to invasion of the roots by soil fungi and, therefore, to protect the tree from root diseases. Most ectomycorrhizal associations in trees are caused by basidiomycetes in the mushroom group (Agaricales) or the puffball group (Gasteromycetes). Some of the species that form ectomycorrhizae include alder, beech, birch, eucalyptus, fir, hickory, larch, pine, oak, and spruce. Vesicular-arbuscular (VA) mycorrhizae do not modify the feeder root anatomy and are not as easily detected as ectomycorrhizae. VA mycorrhizal roots have intracellular hyphae in the root cortex, and the invading fungus often produces vesicles and specialized branched haustoria (arbuscules) in this tissue (Fig. 13.15). Most VA mycorrhizal associations are caused by phycomycetous fungi in the genus *Endogone*. Some of the species that form VA mycorrhizae include almond, apple, ash, avocado, azalea, boxwood, camellia, citrus, dogwood, holly, maple, palms, plum, sycamore, sweetgum, and walnut.

Selected References

Bryan, W. C., and Kormanik, P. P. (1977). Mycorrhizae benefit survival and growth of sweetgum seedlings in the nursery. *South. J. Appl. For.* **1,** 21–23.

Gerdemann, J. W. (1974). Vesicular-arbuscular mycorrhizae. *In* "The Development and Function of Roots: Third Cabot Symposium" (J. G. Torrey and D. T. Clarkson, eds.), pp. 575–591. Academic Press, New York.

Marx, D. H. (1972). Ectomycorrhizae as biological deterrents to pathogenic root infections. *Annu. Rev. Phytopathol.* **10,** 429–454.

Marx, D. H., and Bryan, W. C. (1975). The significance of mycorrhizae to forest trees. *In* "Forest Soils and Forest Land Management" (B. Bernier and C. H. Winget, eds.), pp. 107–117. Presses Univ. Laval, Quebec.

Spitko, R. A., Tattar, T. A., and Rohde, R. A. (1978). Incidence and condition of vesicular-arbuscular mycorrhizae infections in the roots of sugar maple in relation to maple decline. *Can. J. For. Res.* **8,** 375–379.

Trappe, J. M. (1977). Selection of fungi for ectomycorrhizal inoculation in nurseries. *Annu. Rev. Phytopathol.* **15,** 203–222.

14

Discoloration and Decay in Living Trees

INTRODUCTION

A wound is the starting point for interactions between the living tree and the invading microorganisms which can lead to discoloration and decay. Wounds are common events in the life of a tree. Every time a branch is lost or there is a break in the bark to the xylem a wound occurs. Extensive research by Dr. Alex L. Shigo of the U.S. Forest Service has shown that the best way to understand and to prevent discoloration and decay in trees is to study (1) wounds, (2) host responses to wounds, and (3) successions of microorganisms that invade wounds.

WOUNDS

Wounds that can lead to discoloration and decay can occur on any part of the branches, trunk, or roots. Wounds vary in severity. Most are small, close quickly, and go unnoticed, but often wounds are large and take many years to close. The area of xylem exposed from a wound and the amount of xylem disruption occurring beneath the bark are two factors that determine wound severity. Broken branches and mechanical injuries are the most common types of wounds on trees.

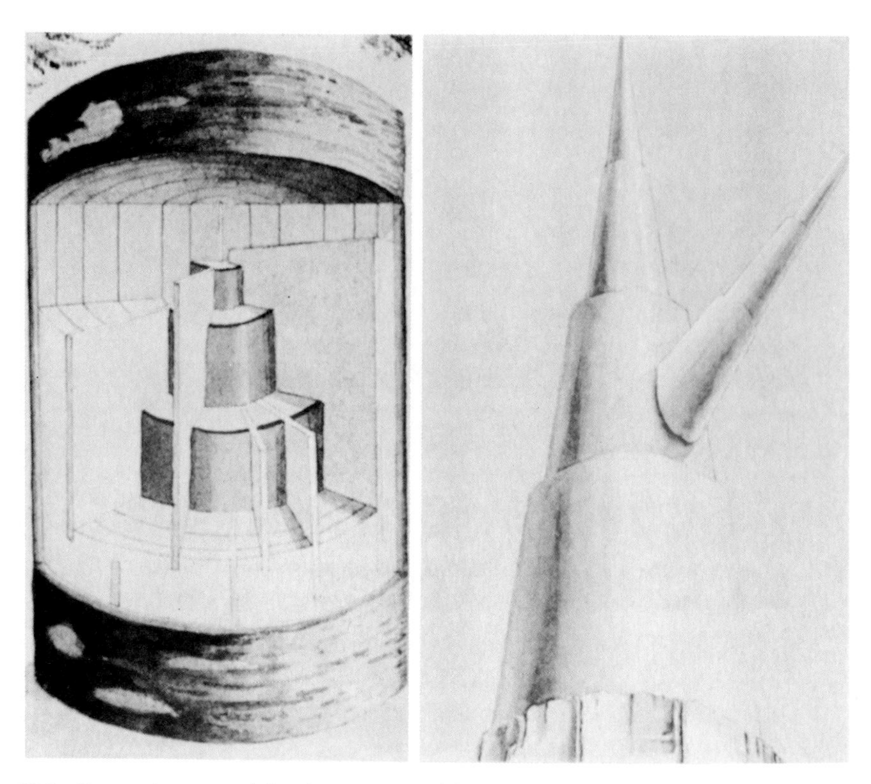

Fig. 14.1 Compartment models of tree stems. (Photographs courtesy of Alex L. Shigo, USDA Forest Service, Durham, New Hampshire.)

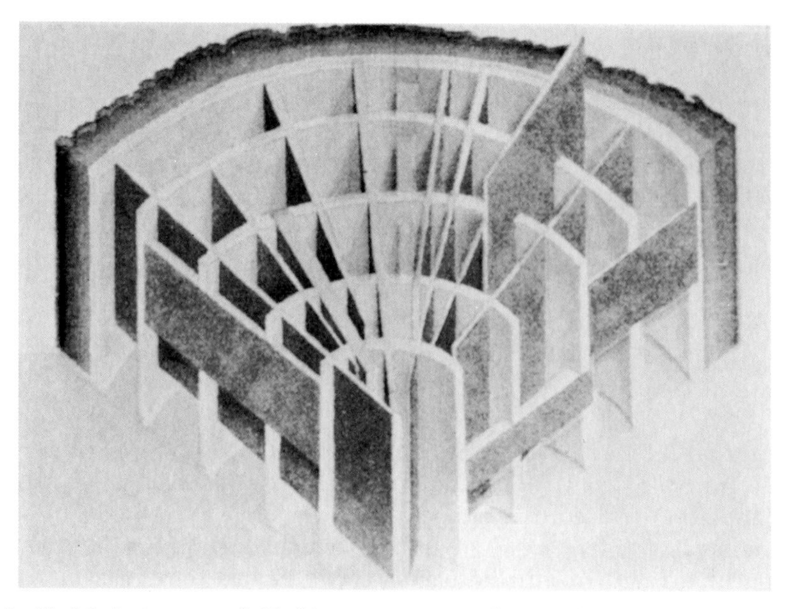

Fig. 14.2 Model of a tree stem divided into compartments by growth rings and rays. (Photograph courtesy of Alex L. Shigo, USDA Forest Service, Durham, New Hampshire.)

HOST RESPONSE TO WOUNDS (CODIT)

A tree responds to all wounds by putting up barriers, both chemical and physical, to block the invasion of wound microorganisms and to confine the infected tissue to as small an area as possible. This process is known as com-

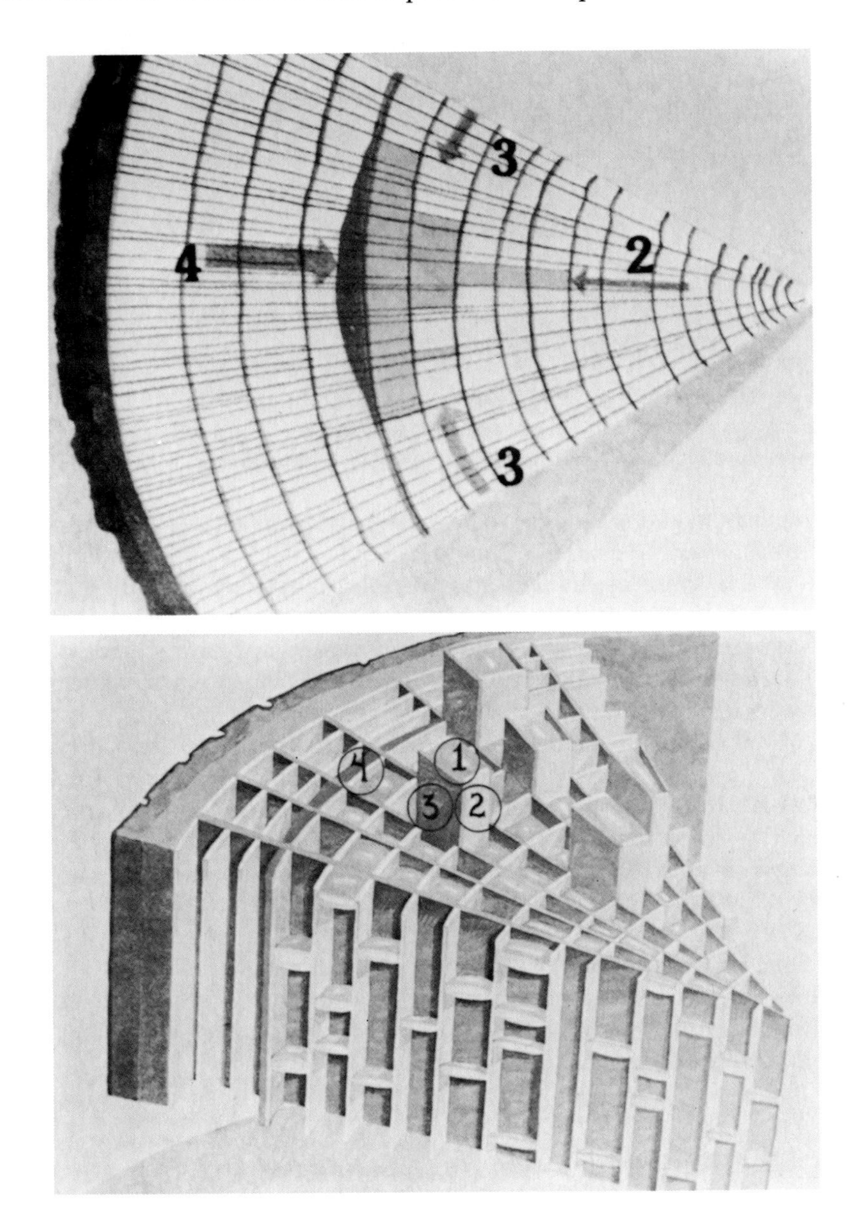

Fig. 14.3 Wall models of a tree's response to wounding. (Photographs courtesy of Alex L. Shigo, USDA Forest Service, Durham, New Hampshire.)

Fig. 14.4 Barrier zone formation (Wall #4) prevents invasion of tissues formed after wounding. (Photograph courtesy of Alex L. Shigo, USDA Forest Service, Durham, New Hampshire.)

partmentalization and is thought to be a major factor in a tree's ability to survive for hundreds, even thousands of years. A tree is a highly compartmented plant (Fig. 14.1). The xylem is divided into annual growth rings radiating outward, which are then divided radially by the rays (Fig. 14.2). When a tree is wounded, barriers to vertical invasion also occur as the vessels and/or tracheids plug.

Compartmentalization of decay in trees (CODIT) is a concept developed by Dr. Shigo which helps in understanding barrier formation by naming each component of the compartment as a wall. Plugging of the vessels and tracheids in the vertical direction forms Wall #1 (Fig. 14.3). The last cells in a growth ring form Wall #2. Ray cells from Wall #3. After a wound, the cambium forms Wall #4, which is called the barrier zone, and it separates xylem formed prior to wounding from xylem formed after wounding. As soon as the tree is wounded, Walls #1–4 begin to form. Wall #1 is the weakest barrier, consequently columns of discolored and decayed wood are often elongate and narrow. Wall #4, the barrier zone, is the strongest wall and prevents microbial invasion into tissues formed by the cambium after wounding. Therefore, once Wall #4 has been formed the column of discolored and decayed wood resulting from that wound cannot exceed the size of the tree when it was wounded (Fig. 14.4).

MICROBIAL INVASION OF WOUNDS

As soon as the tree is wounded, microorganisms on the bark and in the air contaminate the wound surface. Many are able to colonize the wood within the

Fig. 14.5 Fruiting body of *Fomes fraxinophilus* on stem of white ash. (Photograph courtesy of Alex L. Shigo, USDA Forest Service, Durham, New Hampshire.)

wound and invade despite the production of inhibitory chemicals by the tree. Initially, pioneer microorganisms, principally bacteria and imperfect fungi able to utilize the cell contents of the xylem, are most active. Later, wood decay fungi (Basidiomycetes; Hymenomycetes) with the ability to degrade the cellulose and lignin of the cell walls become active (Fig. 14.5). This sequence of pioneer microorganisms followed by wood decay-fungi is known as a microbial succession. It is common in wounds on all living trees and enables microorganisms of different metabolic capabilities to more effectively invade a tree. Clearly it is the invasion of pioneer microorganisms that initially break down the host defenses and invade the xylem that is most critical in the processes that lead to decay.

Selected References

Shigo, A. L. (1967). Successions of organisms in discoloration and decay of wood. *Int. Rev. For. Res.* **2,** 237–299.

Shigo, A. L. (1977). Compartmentalization of decay in trees. *Agric. Inf. Bull. (U.S. Dep. Agric.)* No. 405.

Shigo, A. L. (1979). Tree decay—An expanded concept. *Agric. Inf. Bull. (U.S. Dep. Agric.)* No. 419.

Shigo, A. L., and Hillis, W. E. (1973). Heartwood, discolored wood, and microorganisms in living trees. *Annu. Rev. Phytopathol.* **11,** 197–222.

Shigo, A. L., and Larson, E. vH. (1969). A photo guide to the patterns of discoloration and decay in living northern hardwood trees. *USDA For. Serv. Res. Pap. NE* **NE-127.**

15

Bacterial Diseases

INTRODUCTION

The bacteria have many characteristics that are plantlike. Traditionally they have been placed in the class Schizomycetes of the Thallophyta. Not all taxonomists agree with this placement, and some feel the bacteria belong in a separate kingdom with other prokaryotic organisms.

Bacteria are microscopic, unicellular microorganisms with relatively simple cellular organization. They have a cell wall which is often surrounded by a gelatinous sheath. Inside the wall is the cytoplasm, limited by a plasma membrane, and containing one or more pronuclei. Each pronucleus consists of a single chromosome, compacted into a somewhat globose mass without a nuclear envelope. There are no mitochondria or chloroplasts. Motility is possible in some species by the action of one to many flagella. Reproduction is by simple cellular division, called binary fission. Although very small ($0.2–3 \times 1–6\ \mu m$), each cell can reproduce every 20–30 minutes. It is possible, therefore, for one bacterial cell to yield about 300 trillion bacterial cells in 24 hours. This ability for rapid reproduction is significant in determining the success bacteria have as plant pathogens.

Bacteria are found more commonly in animals than in plants, and are responsible for many serious diseases in both groups. There are 10 orders of bacteria, but only 2 of these (Eubacteriales and Pseudomonadales) contain known plant pathogens. Of these two orders, five genera contain the most important species of plant pathogens: *Corynebacterium, Erwinia, Agrobacterium, Pseudomonas,* and *Xanthomonas.* Most are rod-shaped and none produce spores.

Bacteria are typically not obligate parasites and are unable to penetrate plant tissues directly. They must enter through natural openings such as stomata and lenticels or through wounds. Many species can survive saprophytically in plant refuse, and are instrumental in recycling nutrients for use by other organisms. Method of transmission is generally by insects and rain splash.

To be pathogenic to plants bacteria must multiply and as they multiply they require nourishment. This nourishment must be in soluble form to be assimilated. Rather than actively invading intact cells, bacteria are dependent on their ability to kill and make soluble the host tissue by enzymatic action. By-products of metabolism can also be toxic to vital tissues. Either or both processes will result in various symptoms: necrotic spots, blights, wilts, galls, and cankers.

Because of their small size, bacteria are not easily identified on the basis of colony characteristics or cell morphology. Other means are required; these include utilization of starch, liquefaction of gelatin, production of acid or gas, susceptibility to antibiotics, and stain retention. Many bacteria are host specific and characterized by the types of symptoms they cause, allowing for tentative identification by gross morphological examination.

The impact of bacteria on tree health is probably underrated. Relatively few diseases are known and fewer have been adequately studied. Isolations from healthy as well as discolored and decayed sapwood invariably yield bacteria. This suggests that there may be an attending microflora associated with trees, analogous to that in the digestive tract of human beings and other animals. Further research is required to demonstrate the significance of bacteria in relation to tree pathology.

Selected Reference

Schaad, N. W., ed. (1980). "Laboratory Guide for Identification of Plant Pathogenic Bacteria." Am. Phytopath. Soc., St. Paul, Minnesota.

DISEASE: CROWN GALL

Primary causal agent: *Agrobacterium tumefaciens* (E. F. Smith & Town.) Conn.

Hosts: A large number of both woody and herbaceous plants

Symptomatology: Irregularly shaped galls of various sizes appear on the lower stem (Fig. 15.1), at or near the ground line, and on roots. On some species, the galls may develop higher on the stem and on branches (Fig. 15.2). Growth of

Fig. 15.1 Galls on lower stem of willow induced by *Agrobacterium tumefaciens.* (Photograph courtesy of Shade Tree Laboratories, University of Massachusetts, Amherst.)

Fig. 15.2 Galls in upper branches of willow induced by *Agrobacterium tumefaciens*. (Photograph courtesy of Shade Tree Laboratories, University of Massachusetts, Amherst.)

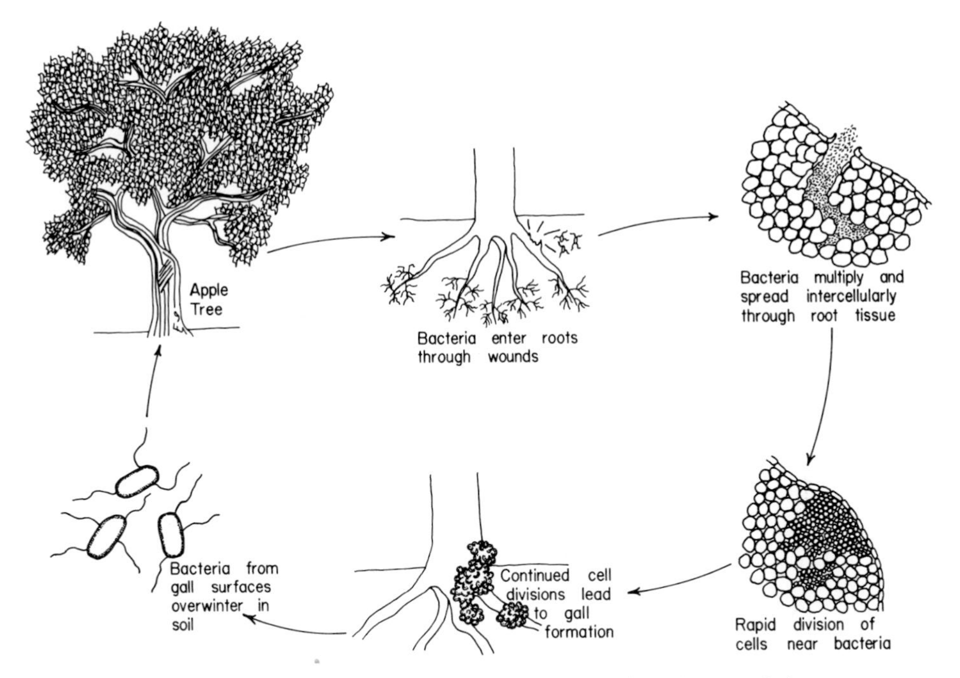

Fig. 15.3 Disease cycle of crown gall caused *Agrobacterium tumefaciens.*

the host tree, especially a young tree in the nursery, is stunted as energy is channeled into gall formation.

Etiology: *Agrobacterium tumefaciens* is a bacterium which overwinters in the soil or in old galls (Fig. 15.3). It enters the host through wounds. Multiplication occurs and the pathogen spreads intercellularly. Crown gall has been likened to cancer in animals because of the way it develops. The bacteria do not contribute much to the structure of a gall, but are responsible for its formation. Host cells adjacent to bacterial cells are stimulated and divide rapidly. This rapid division of cells (hyperplasia) accompanied by enlargement of cells (hypertrophy) results in the formation of a gall. The bacteria are located near the surface of the galls and reenter the soil, where they overwinter.

Control: Since the disease is most prevalent on young nursery stock, care should be exercised to prevent wounds resulting from planting and cultivation procedures. Remove and destroy infected plants in nurseries. Application of bacteriacides to prevent gall formation has been successful, while treatment of existing galls has not been effective.

Selected References

Braun, A. C. (1952). Plant cancer. *Sci. Am.* **186,** 66–72.
Hock, W. C. (1971). Crown gall of woody plants. *Weeds, Trees,* and *Turf* June, pp. 20–21.

Lippincott, J. A., and Lippincott, B. B. (1975). The genus *Agrobacterium* and plant tumorigenesis. *Annu. Rev. Microbiol.* **29,** 377–405.
Moller, W. J., and Schroth, M. H. (1976). Biological control of crown gall. *Calif. Agric.* **30,** 8–9.

DISEASE: FIRE BLIGHT

Primary causal agent: *Erwinia amylovora* (Burrill) Winslow et al.

Hosts: Many species of the rose family, but most severe on apple (*Malus* spp.), pear (*Pyrus* spp.), hawthorn (*Crataegus* spp.), and mountain ash (*Sorbus* spp.)

Symptomatology: Symptoms are first noticeable on succulent growth in late spring. Infected flowers, leaves, and shoots suddenly shrivel and turn dark brown or black as though burned (Figs. 15.4, 15.5). Dead leaves remain at-

Fig. 15.4 Fireblight on apple trees showing blackened leaves and dieback. (Photographs courtesy of William E. MacHardy, University of New Hampshire, Durham.)

Fig. 15.5 Fireblight on mountain ash.

tached to twigs. Cankers develop and eventually girdle larger branches at the base of infected twigs (Fig. 15.6). Oozing pustules containing millions of bacterial cells appear around canker margins.

Etiology: *Erwinia amylovora* is a bacterium that overwinters in infected branches near the margins of cankers (Fig. 15.7). During the spring the pathogen begins to multiply and bacterial ooze forms near the margins of infected tissues. The ooze attracts several species of insects that carry the bacteria to blossoms and leaves. Penetration of the host is gained through nectarthodes of blossoms, through wounds and stomata of leaves, and through wounds in branches. Colonization proceeds intercellularly. Infected blossoms become the source of secondary inoculum which may be spread by rainsplash and insects. As the pathogen multiplies, infected blossoms and leaves shrivel and turn dark brown or black. The bacteria then move into small twigs, which also discolor. Eventually, the bacteria reach large branches where cankers are formed. As

Fig. 15.6 Fireblight cankers on apple stems. (Photographs courtesy of William E. MacHardy, University of New Hampshire, Durham.)

the cankers enlarge they may girdle the stem. The pathogen overwinters in the apparently healthy tissues adjacent to dead areas of the cankers.

Control: Infected parts should be removed by pruning. Pruning tools should be sterilized between cuts to prevent spread of the pathogen. Use of systemic antibiotics in conjunction with plant growth regulating substances have been used with some success. Since the pathogen develops most rapidly in succulent new tissues, the use of fertilizers high in nitrogen, such as calcium nitrate or ammonium nitrate, should be avoided.

Selected References

Baker, U.F. (1971). Fire blight of pome fruits: The genesis of the concept that bacteria can be pathogenic to plants. *Hilgardia* **40,** 603–633.

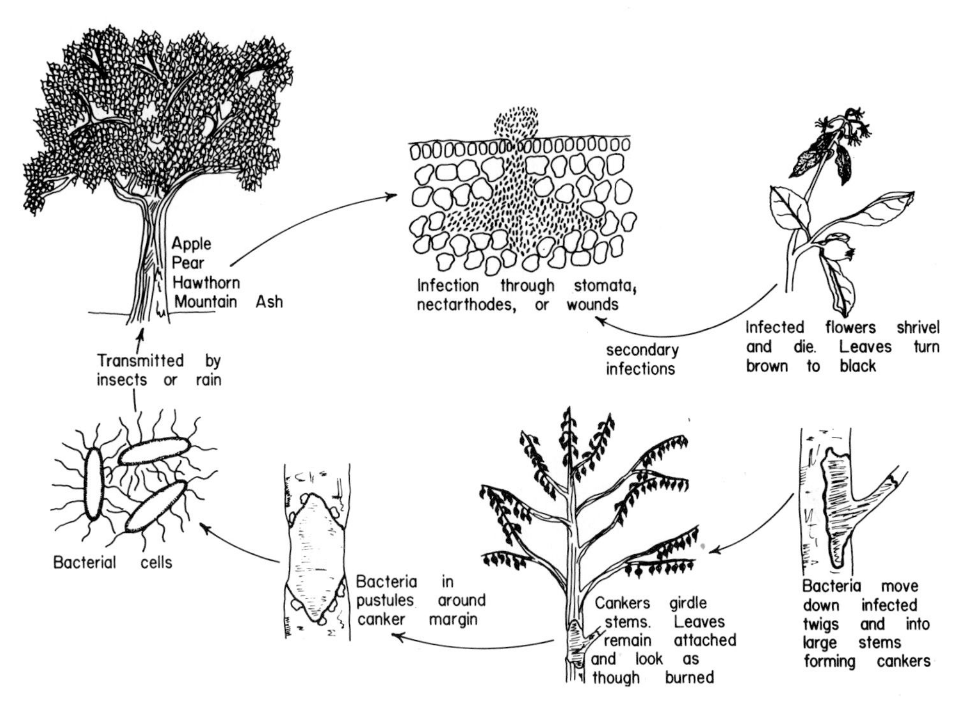

Fig. 15.7 Disease cycle of fireblight caused by *Erwinia amylovora.*

Jones, A. L., and Parker, K. G. (1963). Fire blight blossum infection control on pear with streptomycin, as influenced by adjuvant addition, and with some common fungicides. *Plant Dis. Rep.* **47,** 1074–1078.

Schroth, M. N., Moller, W. J., Thompson, S. V., and Hildebrand, D. C. (1974). Epidemiology and control of fire blight. *Annu. Rev. Phytopathol.* **12,** 389–412.

DISEASE: BACTERIAL WETWOOD OF ELM

Primary causal agent: *Erwinia nimipressuralis* Cart.

Host: American elm (*Ulmus americana* L.)

Symptomatology: The most obvious external evidence of bacterial wetwood in established trees is liquid bleeding from wounds, which is at first colorless to light yellow, turning dark brown upon exposure to the air. The liquid is full of bacteria. Prolonged bleeding from wounds leads to secondary infections by other microorganisms, with the flow of material termed slime flux (Fig. 15.8). Other external evidence includes dieback of the crown and localized mortality of the bark and cambium associated with bleeding wounds. Crossections of infected trees are water-soaked and discolored from the pith outward into the

Fig. 15.8 Slime flux due to prolonged bleeding from wound on American elm.

xylem. In young saplings, the infected pith is olive-green in color, but not visibly water-soaked.

Etiology (Based on Murdoch and Campana from Stipes and Campana, 1981): The disease progresses through four stages of development: (1) Infection-Colonization, (2) Competition-Stabilization, (3) Growth and Development, and (4) Equilibrium (Fig. 15.9). *Erwinia nimipressuralis* and other saprophytic soil bacteria are attracted to young seedlings and presumably infect and colonize the pith when germination of the seed occurs. The role of the saprophytic bacteria and their association with *E. nimipressuralis* is not clear. In the young sapling, the various bacteria compete for survival, eventually leading to relative stabilization of the bacterial populations. Although the pith changes to an olive-green color, it is not water-soaked. This is the prewetwood condition. As the tree matures, the infected tissues become water-soaked. This is the typical wetwood condition. In the stem the wetwood develops in a pattern

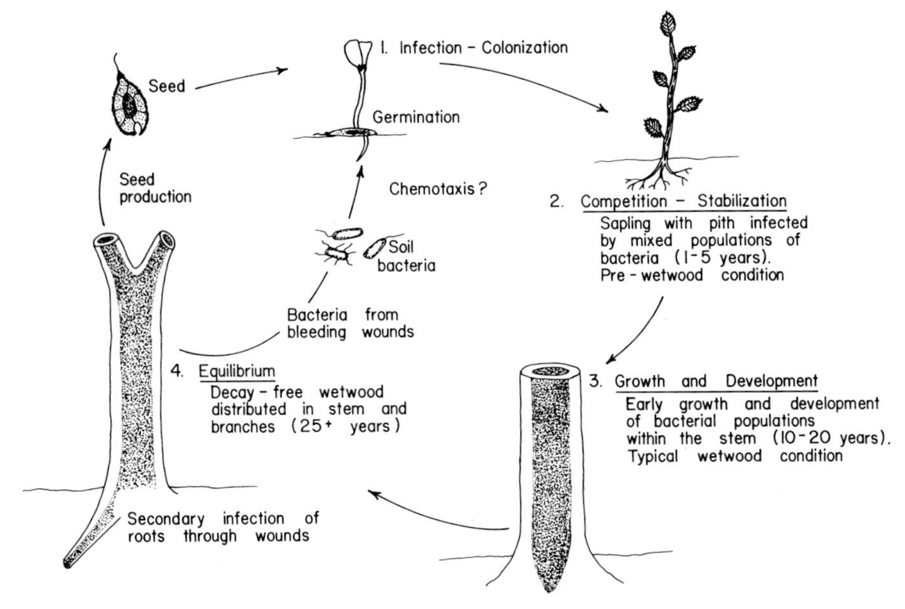

Fig. 15.9 Disease cycle of bacterial wetwood of elm associated with *Erwinia nimipressuralis* and other saphrophytic bacteria. (Redrawn from sketches supplied by C. W. Murdoch and R. J. Campana, University of Maine, Orono.)

similar to an inverted cone. As the pith of each successive branch in the crown is infected and colonized by bacteria, the wetwood enlarges radially in the stem. This pattern continues throughout the life of the tree. In older stems bacterial populations are closely tied to the available food supply and eventually reach equilibrium, with *E. nimipressuralis* predominating in number. Secondary infections occur through wounds in the roots. The wetwood remains decay-free and persists until death of the tree. The pathogen reenters the soil after discharge through bleeding wounds.

Control: Control is probably not feasible nor even desirable, since the disease has not been shown to cause mortality in elm, and the affected wood remains largely decay-free.

Selected References

Campana, R. J., Murdoch, C. W., and Andersen, J. L. (1980). Increased development of bacterial wetwood associated with injection holes made for control of Dutch elm disease. *Phytopathology* **70,** 460. (Abstr.)

Murdoch, C. W., Biermann, C. J., and Campana, R. J. (1980). Pressure and composition of intrastem gases produced in wetwood infected American elm (*Ulmus americana* L.) trees. *Phytopathology* **70,** 466. (Abstr.)

Stipes, R. J., and Campana, R. J., eds. (1981). "A Compendium of Elm Diseases." Am. Phytopathol. Soc., St. Paul, Minnesota.

16

Virus Diseases

INTRODUCTION

Viruses, which are often termed virus particles, are too small to be seen with a light microscope but can be seen with an electron microscope. Most plant viruses consist of a protein coat and of the nucleic acid, RNA (ribonucleic acid), but a few contain DNA (deoxyribonucleic acid) instead of RNA. Their shapes are variable but most are either a rounded form (polyhedral) (Fig. 16.1A) or an elongated form (rod-shaped) (Fig. 16.1B) .Virus particles do not possess any structures for host penetration and rely on vectors such as insects and nematodes or on fresh wounds to be introduced into the tree. Once inside, the virus is reproduced by the living cells and invades living tissues throughout the tree. Injury to the tree results from infected tissues diverting their metabolism from normal activities to the production of virus particles and to abnormal metabolic activities. Virus-infected cells are not generally killed but produce and release virus particles which further invade the tree. This continues for the life of the tree, since once infected, trees remain so for the rest of their lives.

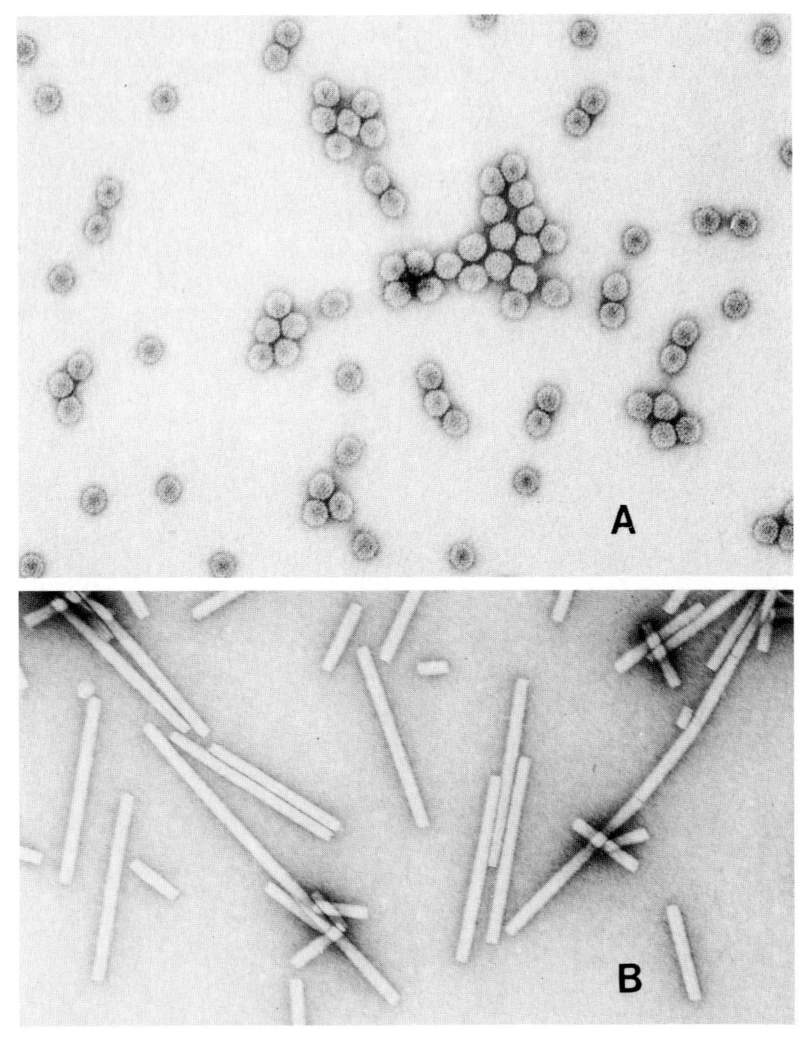

Fig. 16.1 Polyhedral-shaped virus (A), 125,000×, and rod-shaped virus (B), 83,000×. (Photographs courtesy of George N. Agrios, University of Massachusetts, Amherst.)

VIRUS DISEASES OF TREES

Many of the same viruses that infect crop plants such as tomato and tobacco also infect trees. Viruses are spread by vegetative propagation, by nematodes, and by aphids and leafhoppers during feeding. The most common symptoms of virus infection are (1) mottling or mosaic patterns on the leaves, (2) various degrees of stunting of plants, (3) distorted growth, and (4) necrotic lesions. Leaf

Fig. 16.2 Foliar symptoms of virus diseases of trees. Apple mosaic (A), poplar mosaic (B), lilac ringspot (C), birch ring pattern mosaic (D), apple russet ring (E), and European ash TMV (F). (Photographs courtesy of George N. Agrios, University of Massachusetts, Amherst.)

mottles or mosaics, which are the most frequently encountered symptoms, are expressed in a variety of forms in many common virus diseases of trees (Fig. 16.2).

CONTROL OF VIRUS DISEASES OF TREES

Large trees: Virus diseases of pole-sized and larger trees usually cause only minor effects on tree health. In contrast to fruit trees where fruit yield is greatly diminished and sterility is common, virus infections in forest and shade trees often go unnoticed by all but the tree pathologist. However, virus-infected trees usually grow slower than healthy trees and should be discriminated against during thinning, but as shade and ornamental trees they are often attractive and useful. Control efforts on large trees with virus diseases are, therefore, usually not warranted.

Small trees: Vegetative propagation of trees by budding or by rooted cuttings has become very common in many tree nurseries. Using these techniques it is possible to produce trees of known growth and form characteristics. However, trees produced from buds or cuttings taken from infected trees will also carry the viruses into the stock. When seedlings become infected with viruses, they have little chance to become useful forest or shade trees. Care should be taken, therefore, to obtain virus-free plants by selecting only planting stock of known quality. The following measures have been used to control virus diseases in the nursery:

1. Check source plants periodically for presence of viruses by indexing with known indicator plants.
2. If budding or grafting on seedlings, make sure they come from seed obtained from virus-free trees.
3. Keep vector populations in soil or on source plants low with nematocide fumigations and insecticide applications.
4. Heat inactivation—Brief immersion in hot water (43°–57° C) of budwood or the entire plant is effective in certain virus diseases.
5. The recent use of propagation from meristem tip culture and from callus culture holds promise to reduced virus incidence.

Selected References

Agrios, G. N. (1975). Virus and mycoplasma diseases of shade and ornamental trees. *J. Arboricult.* **1,** 41–47.

Fulton, J. P., and Kim, K. S. (1973). A virus resembling tomato spotted wilt virus in black locust. *Plant Dis. Rep.* **57,** 153–155.

Gotlieb, A. R., and Berbee, J. G. (1973). Line pattern of birch caused by apple mosaic virus. *Phytopathology* **63,** 1470–1477.

Hibben, C. R., and Hagar, S. S. (1975). Pathogenicity of an ash isolate of tobacco ringspot virus. *Plant Dis. Rep.* **59,** 57–60.

Hibben, C. R., and Walker, J. T. (1971). Nematode transmission of the ash strain of tobacco ringspot virus. *Plant Dis. Rep.* **55,** 475–478.

Lana, A. O., and Agrios, G. N. (1974). Transmission of a mosaic disease of white ash to woody and herbaceous hosts. *Plant Dis. Rep.* **58,** 536–540.

17

Mycoplasmalike Organism Diseases

INTRODUCTION

Diseases caused by mycoplasmas were once thought to be caused by viruses, but recent studies of their ultrastructure and chemotherapy revealed their different nature. Mycoplasmas, or mycoplasmalike organisms (MLO) as they are often called, bear a close resemblance to bacteria except they lack a cell wall. MLO can assume a number of forms including spheroids, long filaments, and spirals. MLO are usually carried by insect vectors (leafhoppers) and through feeding are introduced into the phloem in which they move throughout the plant. The effects of mycoplasma infection usually include foliar yellowing or reddening, necrosis of vascular tissue, and progressive decline and death, all within one to three seasons. Two important diseases of trees caused by MLO which typify this disease progression are (1) elm yellows and (2) lethal yellowing of coconut palms. Sensitivity to tetracycline antibiotics is another important characteristic of MLO and its use has been quite effective for control of some MLO diseases of trees.

DISEASE: ELM YELLOWS (ELM PHLOEM NECROSIS)

Primary causal agent: Mycoplasmalike organism (MLO)

Vector: White-banded American Leafhopper

Hosts: Native North American elms. European and Asian elms are resistant.

Fig. 17.1 Witches' broom on slippery elm caused by MLO. (Photograph courtesy of Wayne A. Sinclair, Cornell University, Ithaca, New York.)

Symptomatology: Leaf yellowing, droop, and curl are usually the initial symptoms, followed shortly by premature leaf drop in summer. Witches' brooms are sometimes formed on red (slippery) elms (Fig. 17.1). Symptom expression is usually over the entire crown. Severely affected trees generally do not leaf out the following spring. If leaves are produced, they usually are dwarfed and will turn yellow and fall soon after formation. Most often, affected trees die within one year of initial symptom expression, although some trees may persist for several years. Internal symptoms, yellow or butterscotch discoloration of the inner bark, can first be seen as phloem necrosis in the roots. The discoloration is confined to the phloem and only superficially on the outer surface of the xylem (Figs. 17.2, 17.3). It is most easily seen in the field in the lower trunk and buttress. Infected bark of American, cedar, and September

Fig. 17.2 Bark peel of elm stem infected with MLO (right) and healthy elm stem (left). Note discoloration of phloem and outer surface of the xylem (right). (Photographs courtesy of Wayne A. Sinclair, Cornell University, Ithaca, New York.)

Fig. 17.3 Slant cut on elm stem infected with MLO. Note discoloration of innermost phloem and vascular cambial region. (Photograph courtesy of Wayne A. Sinclair, Cornell University, Ithaca, New York.)

elms produces an odor of oil of wintergreen during the summer months. This odor is very helpful in diagnosis since it is only produced in trees with elm yellows.

Etiology: The MLO pathogen is transmitted to susceptible elms during phloem feeding of the vector, the white-banded American leafhopper (Fig. 17.4). The pathogen proliferates in the phloem sieve cells and moves throughout the tree in the phloem. The activities of the pathogen cause phloem necrosis first in the root system and then progressively up the tree into the branches. Noninfected leafhoppers feeding on the leaves can become infected and eventually transmit the disease to healthy susceptible elms. The foliage of the affected elms is killed as the necrosis of the phloem progresses, and the tree dies.

Control: No treatments are effective in curing a tree infected with elm yellows although some temporary symptom remission has been achieved with experimental injection of tetracycline antibiotics. Prompt removal of all newly infected trees can reduce disease incidence in an area.

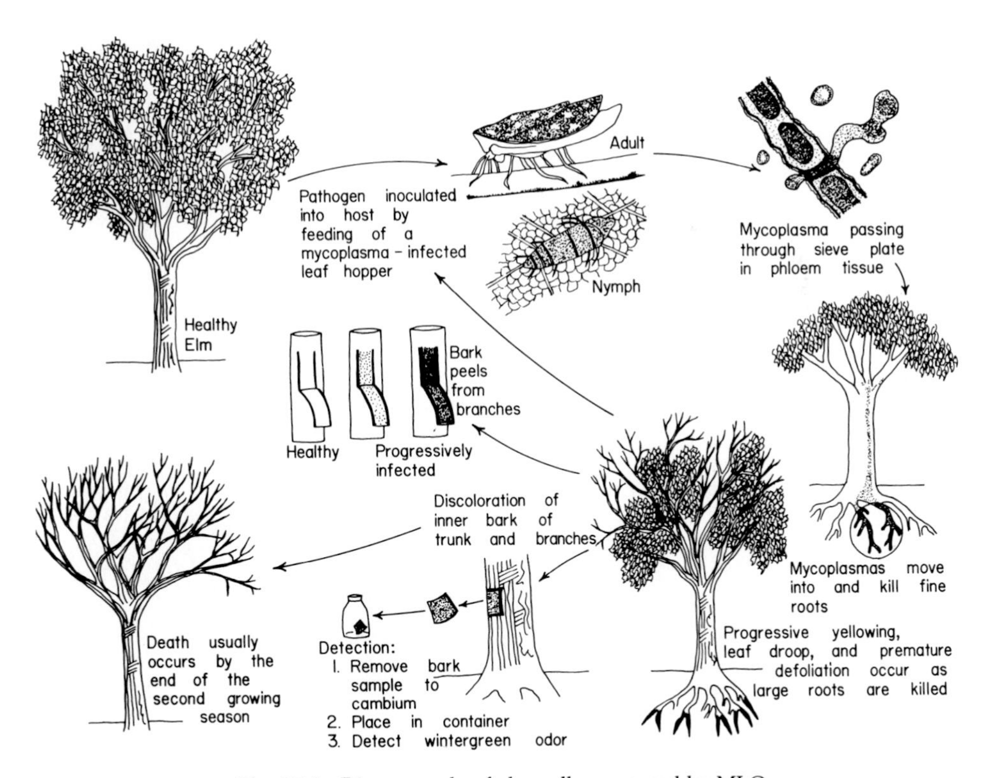

Fig. 17.4 Disease cycle of elm yellows caused by MLO.

Selected References

Baker, W. L. (1949). Notes on the transmission of the virus causing phloem necrosis of American elm, with notes on the biology of its insect vector. *J. Econ. Entomol.* **42,** 729–832.

Braun, E. J., and Sinclair, W. A. (1976). Histopathology of phloem necrosis in *Ulmus americana. Phytopathology* **66,** 598–607.

Filer, T. H. (1973). Suppression of elm phloem necrosis symptoms with tetracycline antibiotics. *Plant Dis. Rep.* **57,** 341–343.

Sinclair, W. A., Braun, E. J., and Larsen, A. O. (1976). Update on phloem necrosis of elms. *J. Arboricult.* **2,** 106–113.

Swingle, R. U. (1939). A phloem necrosis of elm. *Phytopathology* **38,** 757–759.

DISEASE: LETHAL YELLOWING OF COCONUT PALMS

Primary causal agent: Mycoplasmalike organism (MLO)

Vector: Insect vector has not yet been identified.

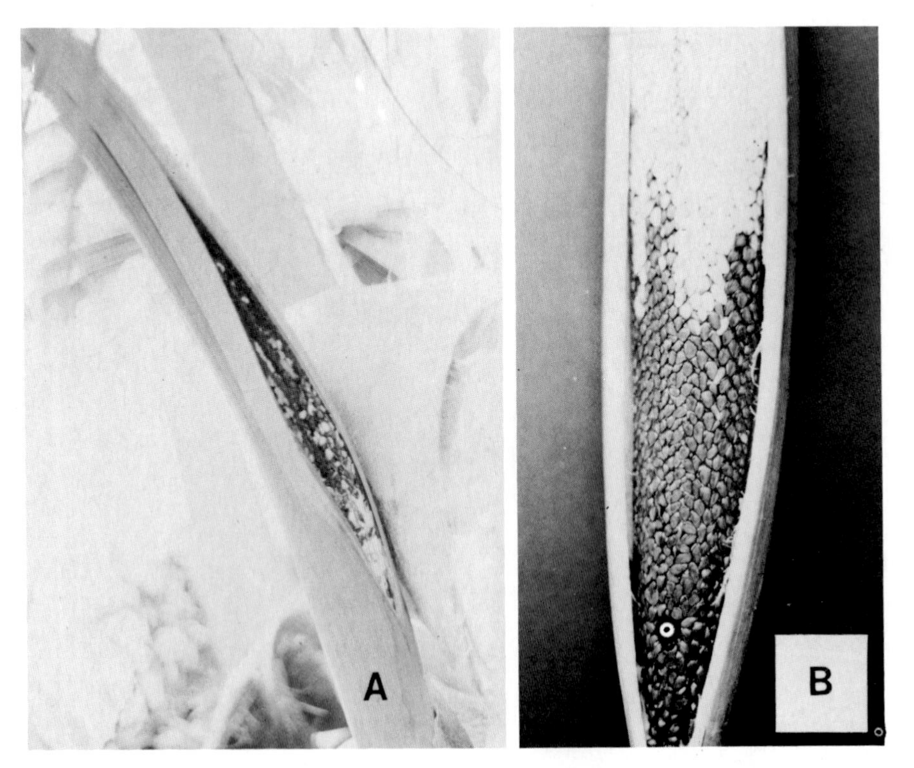

Fig. 17.5 Blackening of inflorescence of coconut palm infected with MLO (A), closeup (B). (Photographs courtesy of Randolph E. McCoy, University of Florida Agricultural Research Center, Ft. Lauderdale.)

Fig. 17.6 Discoloration of fronds of coconut palm infected with MLO. (Photographs courtesy of Randolph E. McCoy, University of Florida Agricultural Research Center, Ft. Lauderdale.)

Hosts: Many species of palm are susceptible. The Malayan dwarf variety of coconut palm is resistant.

Symptomatology: Premature fruit drop is usually the initial symptom followed later by a necrotic blackening of the inflorescences (Fig. 17.5). Discoloration of the lower fronds (leaves) follows (Fig. 17.6), the fronds of some palm species turning golden-yellow while fronds of other species become gray-brown and hang down, often still attached to the tree (Fig. 17.7). Fronds throughout the crown become progressively discolored until death of the terminal growing point occurs and all the leaves soon drop, leaving only the naked trunk.

Fig. 17.7 Drooping fronds still attached to a coconut palm infected with MLO. (Photographs courtesy of Randolph E. McCoy, University of Florida Agricultural Research Center, Ft. Lauderdale.)

Etiology: The lethal yellows MLO is strongly suspected to be transmitted by an as yet unidentified insect vector (Fig. 17.8). Once the MLO is introduced into the phloem it moves through the sieve elements and proliferates. The activity of the pathogen in the phloem disrupts transport through the vascular tissue. As a result, reproductive structures, such as flowers and developing fruit, are killed initially, and then leaves are progressively killed in the final stages. Infected trees are usually dead within three to six months after initial symptom expression.

Control: Avoid planting palm species known to be susceptible to lethal yellowing. Plant resistant varieties such as the Malayan dwarf variety. Apply oxytetracycline antibiotics via trunk injection to susceptible palm species of high value. Injection treatments may be given as a preventative, or therapeutic,

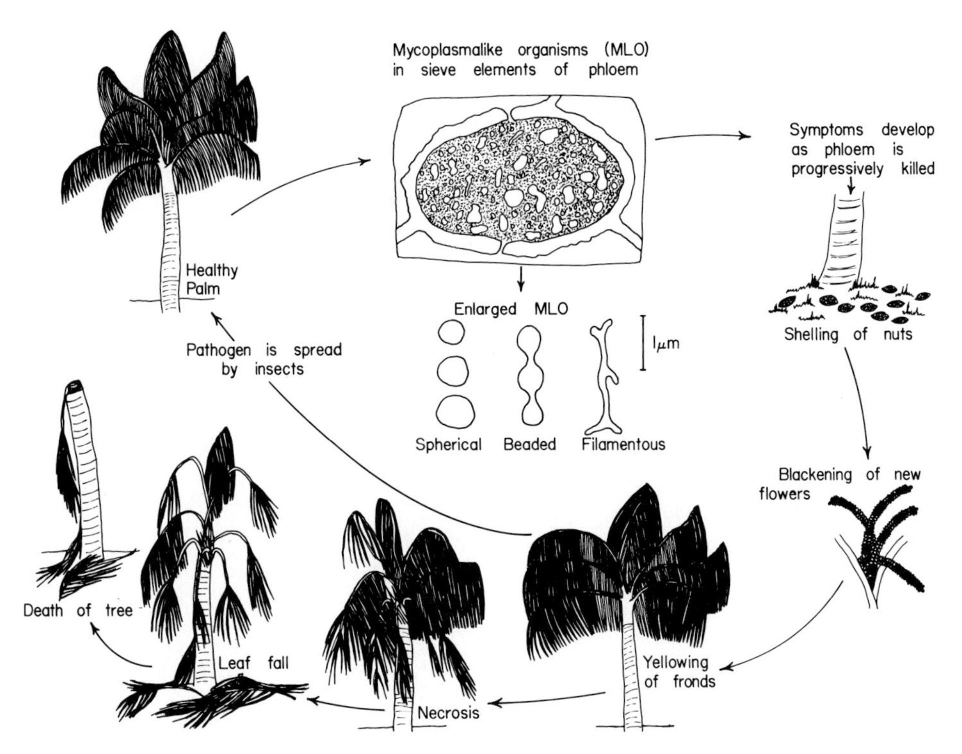

Fig. 17.8 Disease cycle of lethal yellowing of coconut palm caused by MLO.

measure for trees in the earliest stages of the disease. Therapeutic injection treatments, however, must be repeated at approximately 4-month intervals to prevent recurrence of symptoms for the life of the tree. Prompt removal of all infected palm trees is essential to keep disease incidence low.

Selected References

McDonough, J., Zimmermann, M. H. (1979). Effect of lethal yellowing on xylem pressure in coconut palms. *Principles* **23,** 132–137.

Martyn, R. D., and Midcap, J. T. (1975). History, spread, and other palm hosts of lethal yellowing of coconut palms. *Fla. Coop. Ext. Serv., Circ.* No. 405.

Parthasarathy, M. V. (1974). Mycoplasmalike organisms associated with lethal yellowing disease of palms. *Phytopathology* **64,** 667–674.

18

Nematode Diseases

INTRODUCTION

Nematodes are invertebrate animals belonging to the round worms (Phylum Nemathelminthes). Some species of nematodes are parasites of humans and animals and some species are parasites of plants (Fig. 18.1), but most are free-living in the soil. Plant parasitic nematodes possess a hollow daggerlike feeding structure known as a stylet and most are between 0.5 and 5 mm long (Fig. 18.2). Nematodes use the stylet to penetrate plant tissue both to allow feeding on the injured cells and, in some cases, to create an opening for the nematode to move inside the plant. Nematodes can attack any part of a plant, but the most frequent area of attack in woody plants is the root system. The aboveground symptoms of nematode attack are similar to any disease that affects the root system; general growth is poor or stunted and the foliage is often pale green to yellow-green. The symptom pattern is often circular, radiating outward from infection centers.

Life Cycle of Nematodes

The nematode begins life in an egg and undergoes one molt of its exoskeleton or cuticle before it hatches into a motile larva. Nematodes, like insects, must

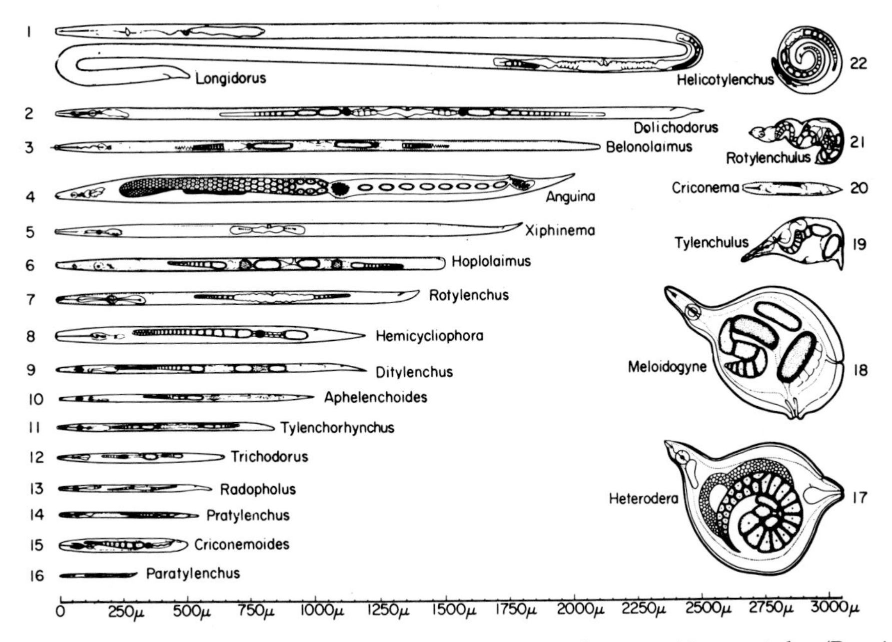

Fig. 18.1 Morphology and relative size of the most important plant-parasitic nematodes. (Drawing courtesy of George N. Agrios, "Plant Pathology," 2nd ed. Academic Press, New York, 1978.)

molt their exoskeletons in order to increase in size. Newly hatched nematode larvae molt their cuticle three more times before they become sexually mature adults. They can feed on plants, however, as soon as they hatch. Adult male and female nematodes mate and a new life cycle begins when the female lays her eggs. Some species, however, have adapted to a parthenogenic life cycle with all individuals being females. Most plant parasitic nematodes have a life cycle which ranges from two weeks to two months. It is, therefore, possible to have a buildup of many generations of nematodes in a few seasons under favorable conditions. The population of plant parasitic nematodes is the most important factor in determining potential for plant injury, since low populations are usually present and only high populations account for substantial injury. Likelihood of injury is highest in seedlings and young trees, especially where repeated plantings can lead to population buildup.

NEMATODE DISEASES OF TREES

Nematode diseases are often categorized by the type of injury caused on the host plant. Belowground injury to roots of trees caused by nematodes usually

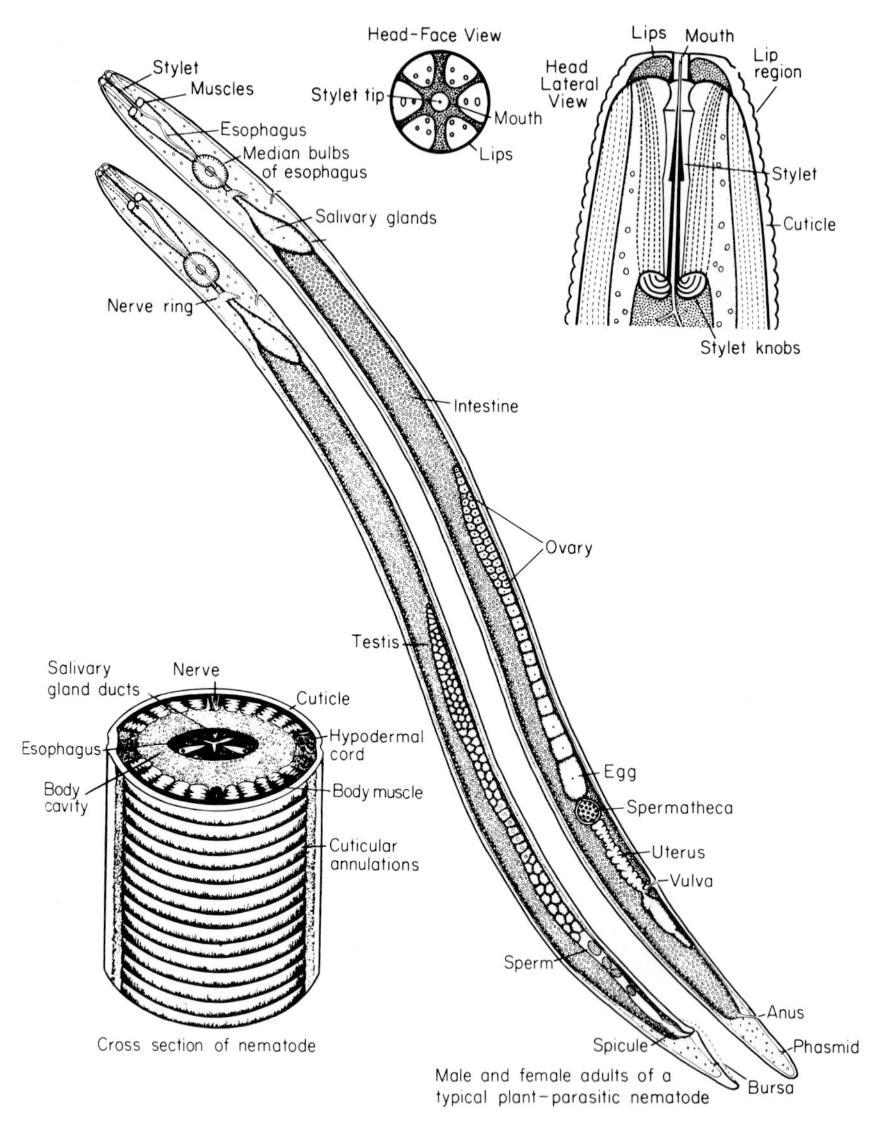

Fig. 18.2 Morphology and main characteristics of typical male and female plant-parasitic nematodes. (Drawing courtesy of George N. Agrios, "Plant Pathology," 2nd ed. Academic Press, New York, 1978.)

Fig. 18.3 Root knot nematode on 12-week-old slash pine seedling. (Photograph courtesy of John L. Ruehle, USDA Forest Service, Athens, Georgia.)

falls into three broad classes: (1) galls or knots (Figs. 18.3, 18.4), (2) lesions (Figs. 18.5, 18.6, 18.7, 18.8), and (3) stubby roots (Fig. 18.9).

Root knot or gall. Swellings form on the root system due to the activity of nematodes inside the root (endoparasitic). Once the nematodes enter the root they spend their whole lives in place (sedentary) drawing nutrition from the stele. Root knot nematodes, *Meloidogyne* and *Meloidodera* spp., are usually responsible for these symptoms.

Root lesion. Necrotic areas develop on smaller roots and the root system in general is lacking in small feeder roots. Injury is caused by these endoparasitic nematodes feeding on and killing tissues while moving through (vagrant) the root cortex. Root lesions are usually caused by the lesion nematodes, *Pratylenchus* spp., the lance nematodes, *Hoplolaimus* spp., and the stunt nematodes, *Tylenchorhynchus* spp.

Stubby roots. The root system is sparse, conspicuously lacking in small feeder roots, but without any noticeable necrosis. Injury is caused by nematodes feeding outside the root (ectoparasitic) and causing the root tip to stop growing. The

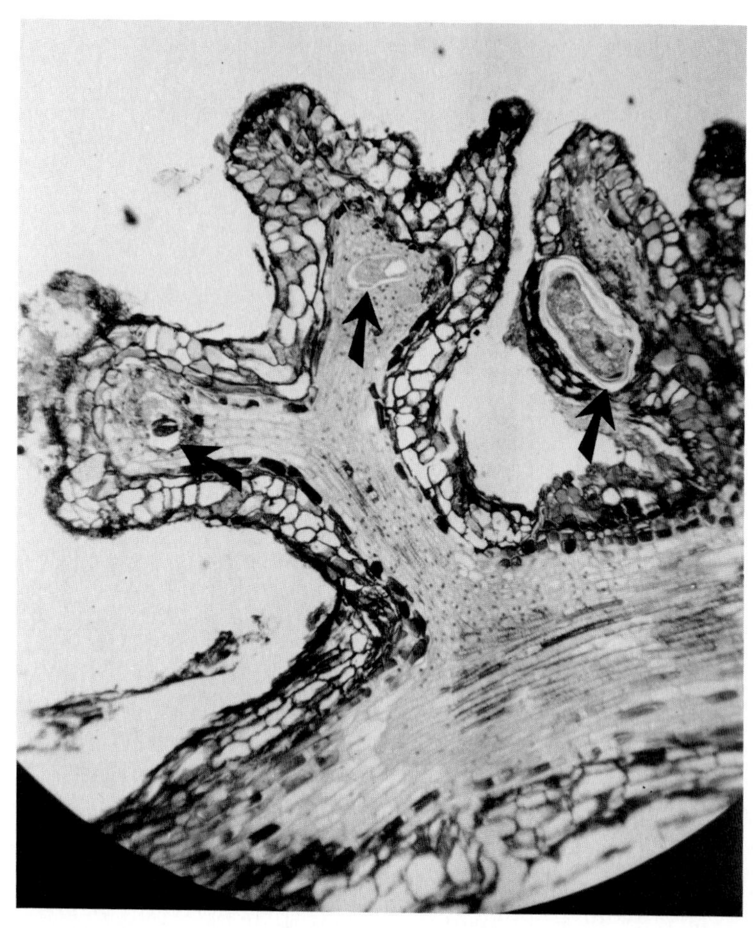

Fig. 18.4 Pine cystoid nematode in ectomycorrhizae of loblolly pine (see arrows). (Photograph courtesy of John L. Ruehle, USDA Forest Service, Athens, Georgia.)

stubby-root nematodes, *Balanolaimus, Paratrichodorus,* and *Trichodorus* spp., and the dagger nematodes, *Xiphinema* spp., are usually responsible for this disease.

CONTROL OF NEMATODE DISEASES

Check roots of all nursery stock to ensure planting of only nematode-free trees. Fumigate soil between plantings in the nursery or prior to planting of any shade or ornamental tree where high nematode populations are present.

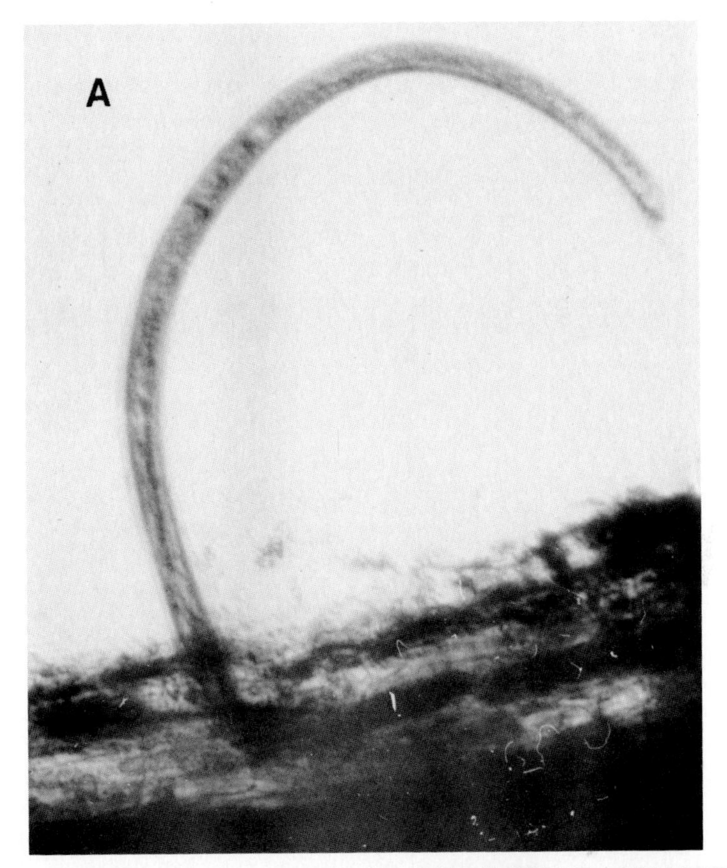

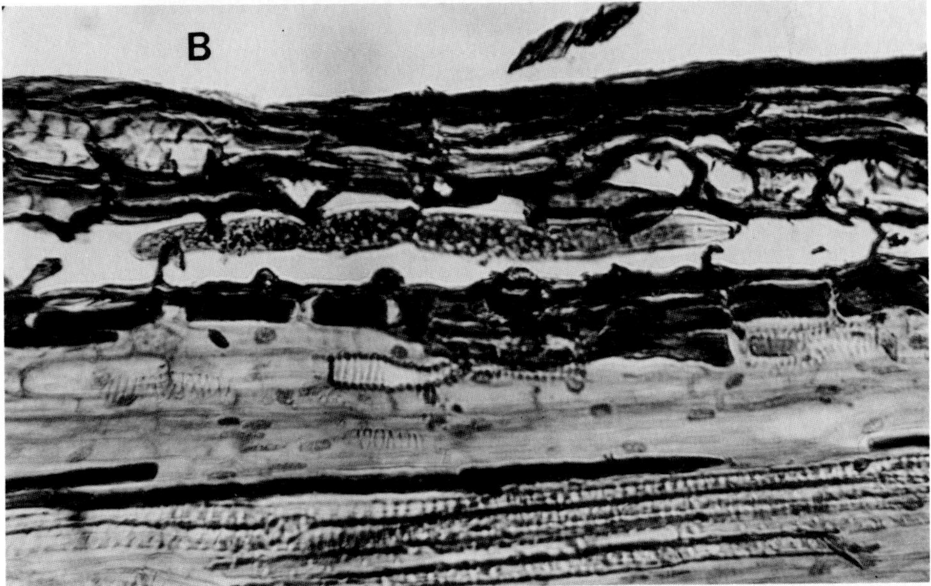

Fig. 18.5 Lance nematode (A) entering the root cortex of loblolly pine (B) inside root cortex. (Photographs courtesy of John L. Ruehle, USDA Forest Service, Athens, Georgia.)

Fig. 18.6 White pine seedlings (1-year-old) in South Carolina nursery area injured by stunt nematodes (A). Healthy (B). (Photographs courtesy of John L. Ruehle, USDA Forest Service, Athens, Georgia.)

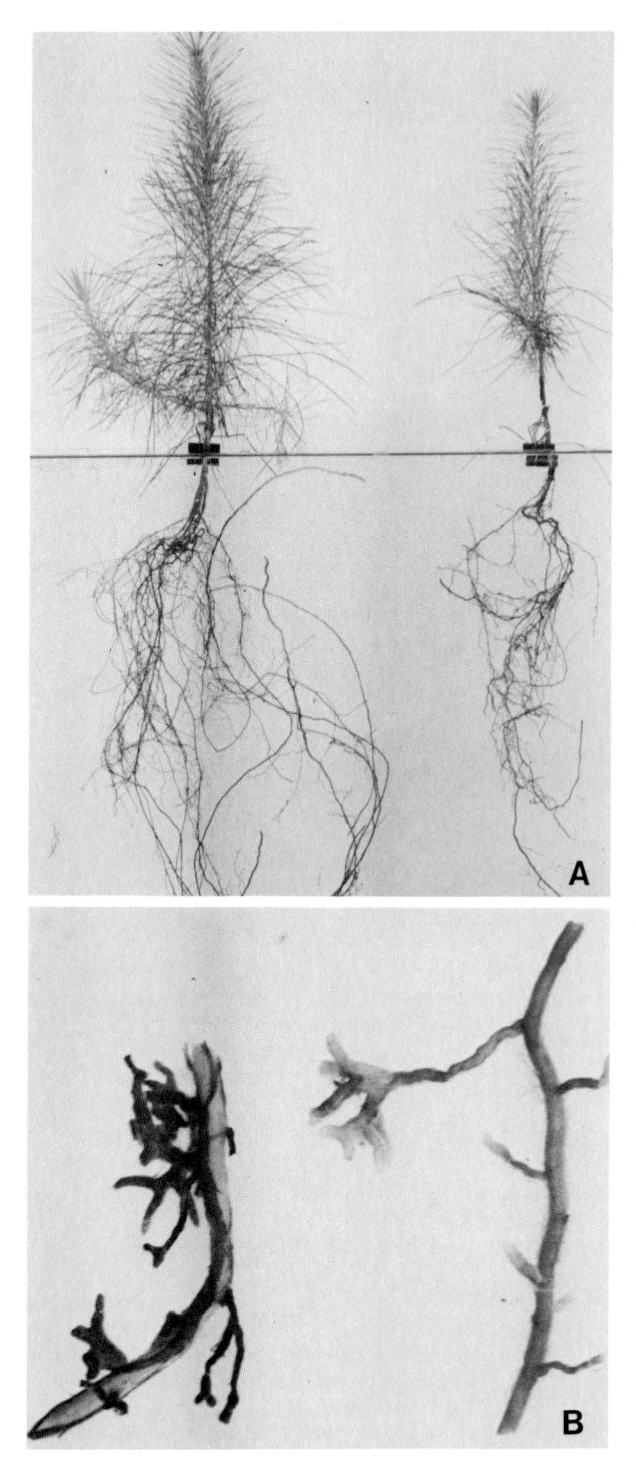

Fig. 18.7 Sand pines. Healthy seedling left and seedling injured by lance nematodes right (A). Closeup of ectomycorrhizal roots (B). Healthy right and injured left. Note darkened roots from root lesion-type injury. (Photographs courtesy of John L. Ruehle, USDA Forest Service, Athens, Georgia.)

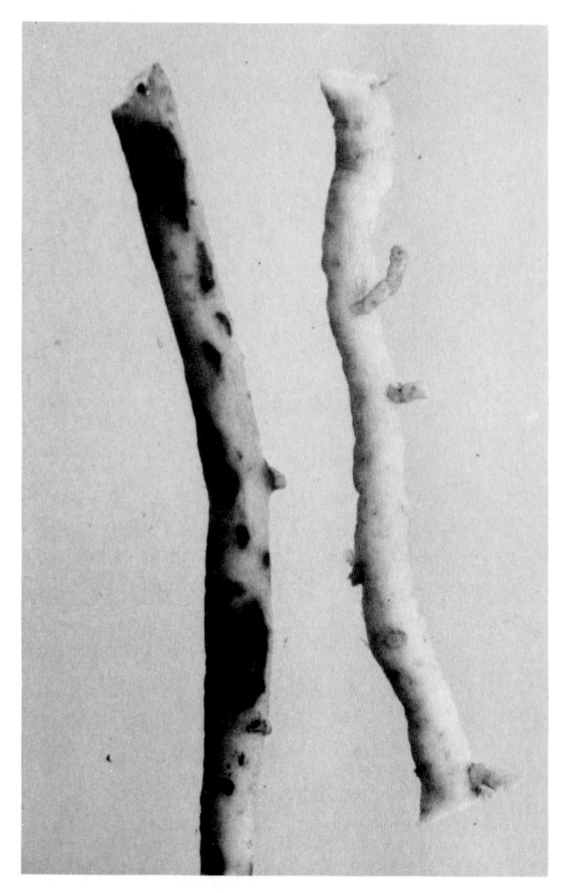

Fig. 18.8 Lesion nematode injury to yellow poplar root segments at left. Note lesions erupting from root surface. Healthy segment at right taken from above the injured segment. (Photograph courtesy of John L. Ruehle, USDA Forest Service, Athens, Georgia.)

Selected References

DiSanzo, C. P., and Rohde, R. A. (1969). *Xiphenema americanum* associated with maple decline in Massachusetts. *Phytopathology* **59,** 279–284.

Ruehle, J. L. (1973). Nematodes and forest trees—Types of damage to tree roots. *Annu. Rev. Phytopathol.* **11,** 99–118.

Ruehle, J. L., and Marx, D. H. (1971). Parasitism of ectomycorrhizae of pine by lance nematode. *For. Sci.* **17,** 31–34.

Sutherland, J. R., and Sluggett, L. J. (1975). Corky root disease: population fluctuations of *Xiphenema bakeri* nematodes, and disease severity in forest nursery soil cropped with different seedling species. *Can. J. For. Res.* **5,** 97–104.

Sutherland, J. R. (1969). Feeding of *Xiphenema bakeri*. *Phytopathology* **59,** 1963–1965.

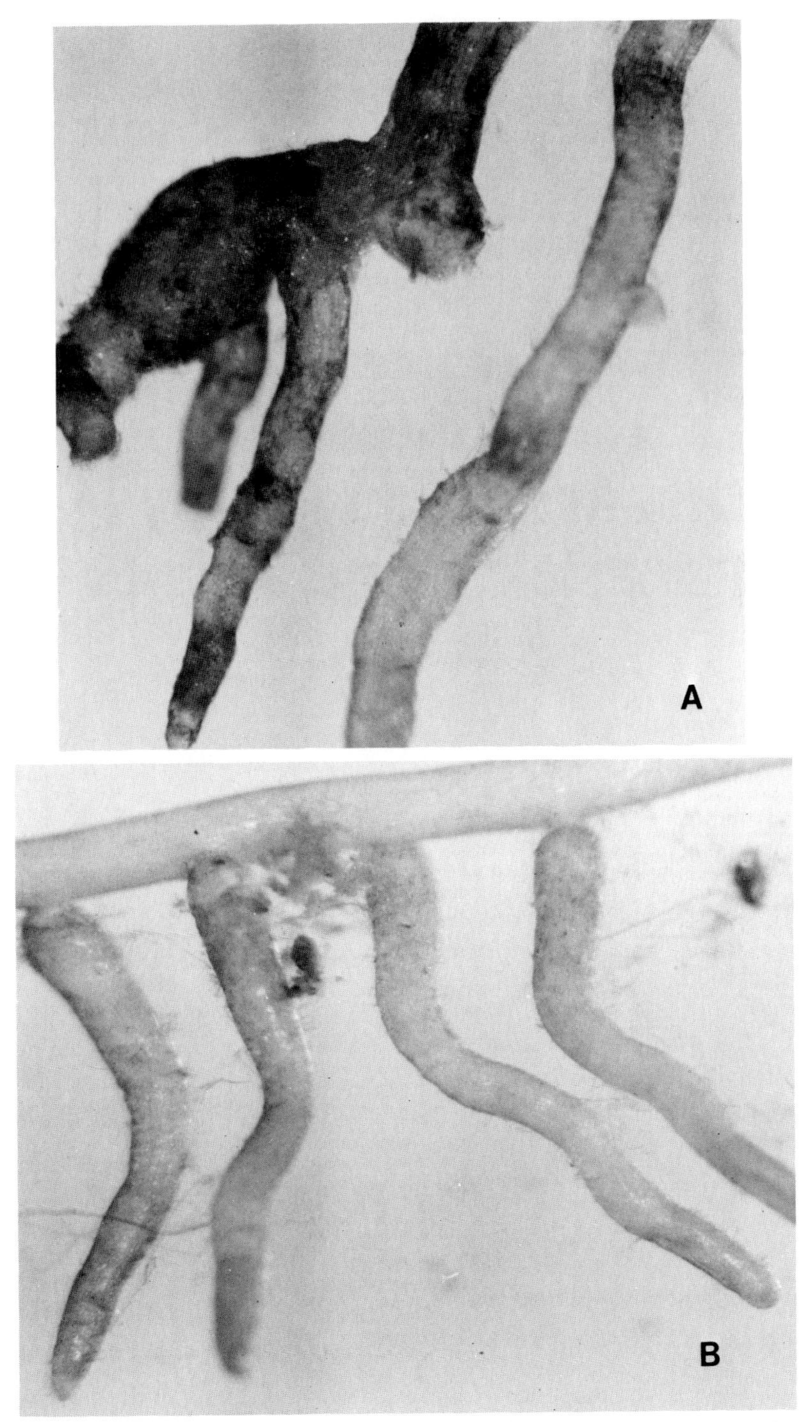

Fig. 18.9 Sting nematode damage (stubby root-type) to sweetgum roots (A). Healthy (B). (Photographs courtesy of John L. Ruehle, USDA Forest Service, Athens, Georgia.)

19

Diseases Caused by Parasitic Seed Plants

INTRODUCTION

Microorganisms such as bacteria and fungi are not the only biotic agents that can cause tree diseases. Higher plants which produce seeds can also be parasitic. Generally, parasitic seed plants lack true roots, true leaves, or both. Specialized rootlike structures penetrate host tissues and extract water and nutrients. Some parasitic seed plants have true leaves and are capable of manufacturing their own carbohydrates through photosynthesis. Of the known 2500 or so species of higher plants that are parasitic on other higher plants, few cause important diseases in trees.

DISEASE: DWARF MISTLETOE

Primary causal agent: *Arceuthobium* spp.

Hosts: Species of *Pinus, Picea, Abies, Larix, Tsuga, Pseudotsuga,* and *Juniperus*

Symptomatology: First evidence of the disease is a swelling of the infected twig at the point of infection. A year or more later, shoots of the parasite emerge from the swollen area (Fig. 19.1). As a result of excessive production of dis-

Fig. 19.1 Shoots of dwarf mistletoe emerging from an infected branch of larch (A) and ponderosa pine (B). (Photograph B courtesy of Frank G. Hawksworth, USDA Forest Service, Fort Collins, Colorado.)

torted host branches, witches' brooms develop. The aerial shoots of the parasite lack true leaves, having primitive structures called bracts. Male and female flowers are produced on separate plants. Female plants are pollinated and produce one-seeded berries. Foliage of infected trees becomes sparse and the upper crown exhibits dieback. Heavily infected trees usually die (Fig. 19.2).

Etiology: Seeds of the dwarf mistletoe plant develop in midsummer to late fall and are forcibly discharged from one-seeded berries for a horizontal distance ranging from 15–30 feet (5–10 m) (Figs. 19.3, 19.4). The seeds are coated with a viscous material and easily adhere to coniferous needles. Rain then washes the seeds down the needles so that they become attached to twigs. Germination generally takes place in late winter or early spring. The germinating radicle produces a holdfast when a suitable penetration point is contacted. The holdfast develops a penetrating wedge of tissue which enters the cortex of the host. Once in the cortex, a network of rootlike structures called the endophytic

Fig. 19.2 Heavily infected ponderosa pine killed by dwarf mistletoe. (Photograph courtesy of Frank G. Hawksworth, USDA Forest Service, Fort Collins, Colorado.)

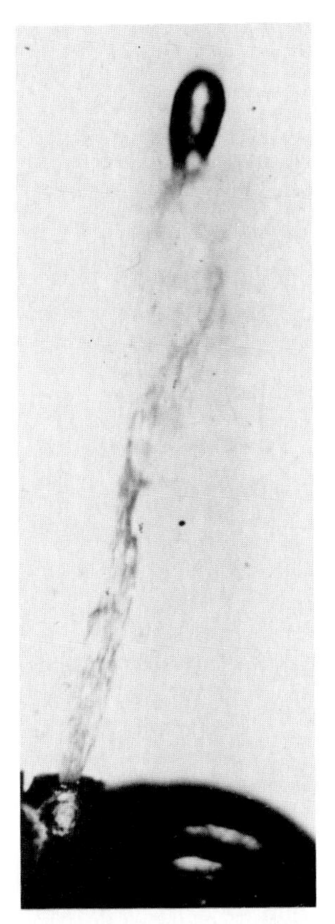

Fig. 19.3 Seed of dwarf mistletoe being forcibly discharged from a one-seeded berry. (Photograph courtesy of Frank G. Hawksworth, USDA Forest Service, Fort Collins, Colorado.)

system ramifies throughout the bark and wood tissues. The endophytic system consists of two parts: (1) the cortical strands which develop in the inner bark, and (2) the sinkers which extend radially from the cortical strands into the xylem (Fig. 19.5). Water and minerals are conducted from host to parasite through the sinkers. About two years after infection, swellings appear around infected tissues. Two to three years later the first shoots of the parasite appear, and mature fruits are produced about two years after shoot emergence. The endophytic system continues to grow as long as host tissues are alive, and new shoots continue to arise. The time from germination to fruit maturity ranges from four to six years, depending on the species of pathogen.

Control: Silvicultural practices are the primary methods of control. Since dwarf mistletoes are obligate parasites, they will live only as long as host tissues do. Therefore, pruning of infected branches or removal of infected trees will render the parasites harmless. Dwarf mistletoes are generally host specific, mak-

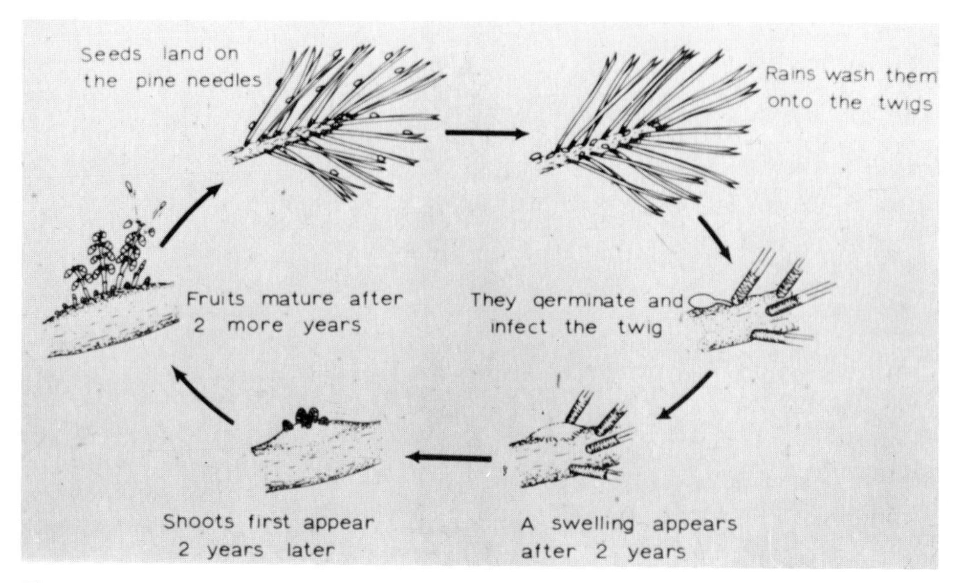

Fig. 19.4 Disease cycle of a typical dwarf mistletoe. (Drawing courtesy of Frank G. Hawksworth, USDA Forest Service, Fort Collins, Colorado.)

Fig. 19.5 Dwarf mistletoe in a pine stem showing shoots (S), cortical strands (C), and sinkers (Si). (Drawing courtesy of Frank G. Hawksworth, USDA Forest Service, Fort Collins, Colorado.)

ing it possible to favor immune or lightly infected species in high disease incidence areas. Chemical and biological controls are of limited value at present, but do show promise for the future.

Selected References

Hawksworth, F. G. (1974). Mistletoes on introduced trees of the world. *U.S. Dep. Agric., Agric. Handb.* No. 469.

Hawksworth, F. G. (1978). Biological factors of dwarf mistletoe in relation to control. *In* Proceedings of the Symposium on Dwarf Mistletoe Control Through Forest Management *U.S. For. Serv., Pac. Southwest For. Range Exp. Stn., Gen. Tech. Rep.* **PSW-31.**

Hawksworth, F. C., and Wiens, D. (1972). Biology and classification of dwarf mistletoes (*Arceuthobium*). *U.S. Dep. Agric., Agric. Handb.* No. 401.

Hudler, G., Nicholls, T., French, D. W., and Warner, G. (1974). Dissemination of seeds of the eastern dwarf mistletoe by birds. *Can. J. For. Res.* **4,** 409–412.

Leonard, O. A., and Hull, R. J. (1965). Translocation relationships in and between mistletoes and their hosts. *Hilgardia* **37,** 115–153.

DISEASE: TRUE (LEAFY) MISTLETOE

Primary causal agent: Species of *Phoradendron* in the United States, species of *Loranthus,* and *Viscum* in other parts of the world.

Hosts: Wide range of hosts, primarily hardwoods growing south of a line across the United States from New Jersey to Oregon. Several species in the West grow on *Juniperus* and *Cupressus* spp.

Symptomatology: The disease is recognized by the presence of mistletoe plants on the trunk and branches of susceptible hosts (Fig. 19.6). Since the mistletoe is an evergreen perennial generally with true leaves, (a few species are leafless), it is most noticeable when leaves of the host have been shed in the fall. Swellings often appear on infected branches. The female plants produce clusters of white to pink, single-seeded berries. Heavy infections cause a slow decline of infected trees.

Etiology: Seeds of the true mistletoe plant are dispersed primarily by birds. Birds feed on the berries and the seeds are passed through the digestive system to be deposited on susceptible hosts. Germ tubes from the seeds penetrate the host and form root systems comparable to those in dwarf mistletoe (see Fig. 19.5). Cortical strands ramify throughout the cortex and wedge-shaped sinkers penetrate into xylem tissues. Since the parasite does produce chlorophyll and consequently carbohydrates, it extracts mostly water and minerals from the host. The root system eventually produces aerial shoots that mature as either male or female plants. Female plants are pollinated and produce clusters of one-seeded berries.

Fig. 19.6 Plants of true mistletoe in the branches of an oak tree.

Control: Because of traditional sentiment associated with true mistletoes, control is seldom practiced or warranted. Selective pruning and dormant application of herbicides have been successful control measures in commercial orchards and in ornamental trees.

Selected References

Hawksworth, F. G. (1974). Mistletoes on introduced trees of the world. *U.S. Dep. Agric., Agric. Handb.* No. 469.

McCartney, W. O., Scharpf, R. F., and Hawksworth, F. G. (1973). Additional hosts of *Viscum album,* European mistletoe, in California. *Plant Dis. Rep.* **57,** 904.

Skelly, J. M. (1972). American mistletoe. *Va. Polytech. Inst. State Univ., Coop. Ext. Serv.* **MR-FTD-19.**

Wagener, W. W. (1957). The limitation of two leafy mistletoes of the genus *Phoradendron* by low temperatures. *Ecology* **38,** 142–145.

DISEASE: DODDER

Primary causal agent: Species of *Cuscuta*

Hosts: Mostly field crops, but also nursery stock of black locust (*Robinia pseudocacia* L.), green ash (*Fraxinus pennsylvanica* Marsh.), and poplar (*Populus* spp.).

Symptomatology: Evidence of the disease first appears as dense tangles of leafless, yellowish strands on the host (Fig. 19.7). During late spring and summer,

Fig. 19.7 Dodder on black locust seedling. (Photograph courtesy of USDA Forest Service.)

the parasite produces clusters of white, pink, or yellowish flowers which then form seeds. Infected hosts are stunted and may be killed.

Etiology: Seeds fall to the ground and overwinter in the soil. During the spring the seeds germinate, producing young, leafless stalks. These stalks "grope" in the air until a host is contacted. When contact is made, the parasite coils around the host stem. Ultimately, haustoria penetrate the host tissues, and the basal portion of the parasite attached to the ground shrivels and dies. The dodder is then completely parasitic on the host. As the parasite matures, flowers and seeds are produced.

Control: Prevention of introduction into nurseries is the best method of control. Use of cleaned seed, soil fumigation before planting, and selective use of herbicides are effective.

Selected References

Gill, L. S. (1953). Broomrapes, dodders, and mistletoes. *In* "Plant Diseases—Yearbook of Agriculture, 1953" (A. Stefferud, ed.), pp. 73–78. U.S. Gov. Print. Off., Washington, D.C.

Hibben, C. R., and Wolanski, B. (1971). Dodder transmission of a mycoplasma from ash witches'-broom. *Phytopathology* **61,** 151–156.

Hocking, D. (1966). *Cuscuta* parasitic on hardwood seedlings. *Plant Dis. Rep.* **50,** 593–594.

Lee, W. O., and Timmons, F. L. (1958). Dodder and its control. *Farmers' Bull.* No. 2117.

PART III

Diseases Caused by Noninfectious Agents

Unlike infectious diseases in which the causal agent is biotic (living), noninfectious diseases are caused by agents that are abiotic (nonliving). The fine line between what should be considered disease and what should be considered injury or disorder has been debated for decades. Many noninfectious agents "walk" this fine line. Is frost damage a disease or an injury? Is mineral deficiency a disease or a disorder? In any case noninfectious agents are tree health problems and are the concern of the tree pathologist.

In this section we will use the terms disease, injury, and disorder where it is convenient to do so, without apology or debate. The confusion can be summarized by the following description: The *disorder* expressed by a *diseased* tree may be the result of chemical *injury*!

Noninfectious diseases of trees are environmentally induced. Whereas people make their own environment, trees are at the mercy of their environment. For example, if people are too cold, they go inside a warm house; if they are too hot, they turn on the air conditioner; if they are hungry, they eat; if they are thirsty, they drink. However, trees do not have these options. Trees grow there they do, because they have evolved to endure most environmental extremes, but in some cases human progress has been too rapid for them to adapt. Noninfectious diseases in trees come from unfavorable environmental conditions from which trees cannot escape. Many of these diseases are difficult to diagnose, because they may actually be the result of stress from more than one agent. Also, noninfectious agents may predispose the tree to diseases caused by infectious agents.

Selected Reference

Levitt, J. (1972). "Responses of Plants to Environmental Stresses." Academic Press, New York.

20

Temperature Extremes

INTRODUCTION

All living things have an optimum growth temperature range. Some grow best at high temperatures such as certain blue green algae which thrive at 180°F (82°C). Others grow best at low temperatures such as certain bacteria and fungi which grow at the north and south poles. The summer temperature of the south side of a pine tree may reach 130°F (54°C), and soil temperatures have been recorded from 130°–170°F (54°–77°C). Trees have evolved to thrive during hot summer days and to survive during cold winter days. But, when temperatures go beyond tolerable limits, or when sudden changes in temperature occur, death of tissues is likely.

HIGH TEMPERATURE—HEAT INJURY

Symptoms of heat injury are related to the nature and source of high temperature stress. Seedlings exposed to excessively high temperatures may collapse as the flow of nutrients and water is restricted. The sudden exposure of previously shaded trees to sunlight can cause scorching of leaves, scalding of bark, and dieback (Fig. 20.1). Fissures in the bark become portals of entry for infectious

Fig. 20.1 Symptoms of heat injury. Marginal scorch of leaf (A), scalding of bark suddenly exposed during house lot clearing (B), and dieback from sudden exposure when large elm trees in foreground were cut in early summer (C).

agents that may cause cankers. Scorching is often found associated with logging operations and building construction which involve clearing of wooded areas. When such operations occur during the summer, normally shaded, more succulent tissues are exposed to conditions for which they have insufficient time to adapt. Whenever possible, tree removal should be done at a time when temperatures are low and trees are not in leaf. The problem becomes more difficult when conifers are involved. Trees next to roadways, buildings, and parking lots can also suffer scorch due to reflected heat from these surfaces.

Heat from fires should also be considered, particularly where understory growth is controlled by prescribed burning. Sublethal temperatures reduce the vigor of trees, predisposing them to infectious agents.

The mechanism of heat injury is not clearly understood, but a partial explanation involves the release of toxic substances due to protein denaturation.

Fig. 20.2 Trunk crack resulting from unequal expansion and contraction of tissues subjected to rapidly changing temperatures.

Fig. 20.3 Spring frost injury (A) and subsequent recovery (B) of frost sensitive trees. (Photographs courtesy of Shade Tree Laboratories, University of Massachusetts, Amherst.)

LOW TEMPERATURE—COLD INJURY

Symptoms from cold injury vary depending on the time of year in which the tree is exposed to damaging temperatures. Continuously cold temperatures in the winter do not affect dormant woody tissues. However, when warm temperatures are followed by rapidly declining temperatures, such as can occur in early spring or late fall, unequal expansion and contraction of tissues may result in trunk cracks (Fig. 20.2). During the spring, as trees emerge from dormancy, frosts can cause severe damage. Below normal temperatures *before* bud-break may cause bud mortality resulting in development of witches' brooms. Frosts *after* bud-break can cause death of blossoms, leaves, and shoots (Fig. 20.3). Summer temperatures are seldom low enough to cause damage although an occasional summer frost has been reported in some northern regions.

As the summer days get shorter, trees undergo metabolic changes which prepare them for winter. Occasionally, some tissues may continue to grow during this time, perhaps as a result of late summer fertilization. Growing tissues are susceptible to fall frosts which can cause twig and branch dieback.

Damage from cold temperatures is generally due to physical injury resulting from the formation of ice crystals in or between cells.

Selected References

Maguire, W. P. (1955). Radiation, surface temperature, and seedling survival. *For. Sci.* **1,** 277–284.

Parker, J. (1963). Cold resistance in woody plants. *Bot. Rev.* **29,** 123–201.

Ronco, F. (1975). Diagnosis: "Sunburned" trees. *J. For.* **73,** 31–35.

Wagener, W. W. (1970). Frost cracks—A common defect in white fir in California. *U.S. For Serv., Pac. Southwest For. Range Exp. Stn., Res. Note* **PSW-209.**

White, W. C., and Weiser, C. J. (1964). The relation of tissue desiccation, extreme cold, and rapid temperature fluctuations to winter injury of American Arborvitae. *Proc. Am. Soc. Hortic. Sci.* **85,** 554–563.

21

Moisture Extremes

INTRODUCTION

Water in trees may contribute as much as 95% to the fresh weight of green tissues. The importance of maintaining this moisture percentage should not be underestimated. Water functions in many ways including maintenance of turgor pressure, as a solvent for the transport of cellular constituents, as a component in the production of carbohydrates and other compounds, and as a type of cooling system. Constituents in the soil are dissolved in water and absorbed by the root system. Too much or too little water will affect the efficiency of the root system to perform its function, resulting in injury to aboveground parts.

WATER DEFICIENCY—DROUGHT

Lack of sufficient water will cause various symptoms in trees, depending on the duration and the extent of the water deficit. Excessive temperatures will also contribute to the amount of injury. Symptoms of insufficient plant moisture often appear first as necrotic lesions at the margins and between the veins of leaves, progressing toward the veins as the deficiency continues (Fig. 21.1). Dieback of twigs and branches may follow (Fig. 21.2). Wilt is commonly as-

Fig. 21.1 Symptoms on sugar maple leaf due to insufficient water.

sociated with severe water deficiency and results from loss of turgor pressure. "Winter drying" of some conifers is actually a drought symptom due to excessive transpiration on warm winter days when frozen roots cannot replace transpired water (Fig. 21.3).

WATER EXCESS—FLOODING

Too much water can also be detrimental to trees. Some species have adapted to grow best under very wet conditions, but others, when suddenly subjected to prolonged flooding, will not survive. Prolonged flooding is usually the result of construction projects (highways, dams, buildings, etc.) which interfere with the normal drainage patterns in an area (Fig. 21.4). The usual symptom of water excess is leaf chlorosis followed by death of the entire tree, if the flooded condition persists. Two theories have been suggested to explain the reasons for these chlorotic symptoms. One hypothesis suggests that flooded soils accumulate substances toxic to trees. Further, the soil becomes anaerobic, with subsequent proliferation of anaerobic microorganisms. These microorganisms may also produce substances that are toxic to trees. The second hypothesis suggests that flooding interferes with the natural and required exchange of gases between the soil and the tree. It is not unreasonable to assume that both hypotheses are correct.

Fig. 21.2 Dieback of sugar maple tree due to insufficient water.

Fig. 21.3 Winter drying of arborvitae. Lighter areas are dead.

Fig. 21.4 Tree mortality due to flooding resulting from highway construction and subsequent interference with normal water drainage.

Selected References

Ahlgren, C. E., and Hanson, R. L. (1967). Some effects of temporary flooding on coniferous trees. *J. For.* **65,** 647–650.

Broadfoot, W. M., and Williston, H. L. (1973). Flooding effects on southern trees. *J. For.* **71,** 584–587.

Henckel, D. A. (1964). Physiology of plants under drought. *Annu. Rev. Plant Physiol.* **15,** 363–386.

Kozlowski, T. T., and Davies, W. J. (1975). Control of water loss in shade trees. *J. Arboricult,* **1,** 81–90.

Rusden, P. L. (1967). Drought—Its effects on trees and what can be done about it. *Proc. Int. Shade Tree Conf.* **43,** 93–105.

Stransky, J. J., and Daniel, H. E. (1964). References on effects of flooding on forest trees. *U.S. For. Serv., Southeast For. Exp. Stn., Res. Notes* No. 12.

22

Nutrient Abnormalities

INTRODUCTION

Most soils contain the correct balance of essential minerals needed by trees growing naturally on them. However, planted trees may not be well adapted to the soil conditions in which they are transplanted. Soil conditions can become modified in ways that make some nutrients unavailable to roots. This results in deficiencies. Soils may also be modified by the addition of high concentrations of minerals around the root zone, which can result in toxicities.

Essential nutrients can be placed into two broad categories: (1) macronutrients, consisting of calcium, magnesium, nitrogen, phosphorus, potassium, and sulfur, and (2) micronutrients, consisting of boron, copper, iron, manganese, molybdenum, and zinc, and possibly other "trace elements." Macronutrients are needed by woody plants in relatively large concentrations, and are usually added in amounts ranging in pounds per inch of stem diameter. On the other hand, micronutrients are only needed in small amounts, usually expressed as parts per million (ppm). If any one of these essential nutrients is present in an abnormally high or low concentration, growth of the plant may suffer, and severe injury can occur.

MACRONUTRIENTS

One of the most common arboricultural practices is fertilization. The common macronutrients in most fertilizers, nitrogen, potassium, and phosphorus are added around tree roots to achieve continued growth and maintain vigor. Annual fertilization is recommended for most shade trees since often they receive insufficient quantities of these essential elements from the environment for ideal growth. Regular fertilization is also essential to continued growth of seedlings in a nursery.

MICRONUTRIENTS

Microelement availability can also be altered by modifications of the soil environment. Common abnormalities involve iron, boron, and copper. Alkaline soils high in calcium often make iron unavailable to the roots of many tree species. Since iron is involved in the production of chlorophyll, the affected leaves appear chlorotic (yellow-green or yellow in color). This "lime-induced" chlorosis can cause the decline and death of susceptible trees if not treated. Complete but temporary recovery can usually be achieved by injecting iron directly into the tree trunk, or by adding chelated iron compounds to the soil. Laundry waste water, accidentally flushed around trees or used for irrigation in arid climates, can cause boron toxicity from borax detergents. Flushing with fresh water can correct the problem if diagnosed in the early stages. On the other hand, large areas in New York and New England have boron-deficient soils, where the addition of a little borax improves tree health. Copper sulfate-treated burlap can cause copper toxicity injury to roots of balled and burlapped nursery stock. Runoff from copper leaders and drainpipes can cause a similar toxicity problem to foundation plantings. After removal of the copper source, application of fertilizer with high phosphate content and regular watering will induce normal root growth and should restore vigor.

SOIL AND TISSUE ANALYSIS

Nutrient abnormalities, like most plant diseases, are best treated in the early stages, but are often difficult to diagnose in the field. Soil and plant tissue analyses are the best aids to early diagnosis and prevention of harmful nutrient stress. Caution is advised in the exclusive use of a soil test because essential elements may be present, but in a form unavailable for root absorption. Soil testing is usually available at state universities, often at no charge. In addition, assistance in making specific treatment recommendations and suggestions for further analyses, if needed, are available at this source or from the State

Cooperative Extension Service. A soil test before planting is a step that should be included in any list for correct tree planting, and could prevent the needless loss of newly planted trees.

Selected References

Himelick, E. B., and Himelick, K. J. (1980). Systemic treatment for chlorotic trees. *J. Arboricult.* **8**, 192–196.

Kielbaso, J. J., and Ottman, K. (1976). Manganese deficiency—Contributory to maple decline. *J. Arboricult.* **2**, 27–32.

Kuhns, L. J., and Sydnor, T. (1975). The effects of copper-treated burlap on balled and burlapped *Cotoneaster divaricata,* Rehd. and Wils. *HortScience* **10**, 613–614.

Neely, D. (1976). Pin oak chlorosis. *J. For.* **71**, 340–342.

Smith, E. M., and Gilliam, C. H. (1980). Sources and symptoms of boron toxicity in container grown woody ornamentals. *J. Arboricult.* **8**, 209–212.

23

People Pressure Diseases

INTRODUCTION

People are the dominant force in the shade tree environment and in forest recreation areas. Detrimental pressures on trees from the activities of people are often of much greater consequence to the tree than the effects of microbial pathogens and harmful insects. People pressure diseases (PPD) are a complex and enlarging group of people-related stresses that commonly affect trees. PPD includes air pollution, chemical injury, construction injury, electrical injury, light pollution, and many more factors. Construction injury and chemical injury are described below to illustrate two common types of PPD.

CONSTRUCTION

Trees grow best in an environment of minimal change. Unfortunately, urban and suburban environments and forest recreation areas are locations where drastic changes often occur around trees. Such changes include driveway (Fig. 23.1) and sidewalk (Fig. 23.2) installation, grade changes, road widening, and sewer trenching (Fig. 23.3). Construction activities near trees constitute a serious encroachment on growing space, a condition that can be fatal (Fig. 23.4). Since nearly all of a tree's important feeder roots are found in the upper 12 inches of

Fig. 23.1 Disturbance of root zone of trees due to driveway construction.

Fig. 23.2 Disturbance of root zone of trees due to sidewalk construction.

Fig. 23.3 Disturbance of root zone of trees due to sewer trenching.

Fig. 23.4 Girdling roots resulting from construction activities. (Photographs courtesy of Shade Tree Laboratories, University of Massachusetts, Amherst.)

Fig. 23.5 Filling over roots, causing suffocation.

soil, even the most minor soil disturbance around the tree can cause substantial root injury. Filling over roots during construction is equally harmful by causing suffocation (Fig. 23.5). Roots need to respire by exchanging the carbon dioxide they produce for oxygen from the air. This exchange is essential for roots to live and to function in absorption and transport. When roots are trapped under impervious fill, even if only a few inches thick, they cannot respire, and they die.

If at all possible, construction should be avoided around valuable trees. Keep as much of the root system as possible undisturbed during unavoidable construction. Any tree exposed to even minor construction injury should be fertilized annually with complete formulations low in nitrogen but high in phosphorus, and should be watered during dry periods.

CHEMICAL INJURY

Trees are constantly being exposed to the chemicals we add to our environment. Some chemicals, such as fertilizers and pesticides, are beneficial to trees, if used properly, but are harmful when used improperly. Other chemicals, such as weed and brush killers, are meant to eradicate unwanted plants but must be used with utmost care near shade trees. In addition, trees, especially those growing along streets, are frequently exposed to deicing salts (Fig. 23.6), underground gas leaks, and a large variety of spills of noxious materials, such as petroleum products and antifreeze.

Fig. 23.6 Salt injury of white pine on downhill side of a salted road.

Chemical injury can be minimized by the proper use of materials designed for tree therapy, and the avoidance of harmful materials around trees. Education and attitude are two key factors needed to overcome this problem. Many people do not realize that, as far as therapeutic chemicals are concerned, the old adage "if a little is good, more is better" does not apply. Twice as much pesticide as recommended will not make the pathogen "twice as dead" but may instead cause injury to the tree. In addition, people do not often associate many common but toxic chemicals with plant injury. For example, careless spillage of crankcase oil, gasoline, chlorinated pool water, and antifreeze can severely injure valuable trees and shrubs. The best therapy for most chemical spills is flushing with fresh water. Activated charcoal worked gently into the soil in the spill area is also effective in inactivating many persistent materials.

Selected References

Dirr, M. A. (1976). Selection of trees for tolerance to salt injury. *J. Arboricult.* **2**, 209–216.

Hibbs, R. (1978). Recognition of weed-killer injury to trees. *J. Arboricult.* **4**, 189–191.

Howe, V. K. (1974). Site changes and root damage: Some problems with oaks. *Morton Arbor. Q.* **10**, 49–53.

Tattar, T. A. (1974). People-pressure diseases of trees. *Coop. Ext. Serv., Univ. Mass.*

Westing, A. H. (1969). Plants and salts in the roadside environment. *Phytopathology* **59**, 1174–1181.

Wilson, C. L. (1977). Emerging tree diseases in urban ecosystems. *J. Arboricult.* **3**, 69–71.

Yingling, E. L., Keeley, C. A., Little, S., and Burtis, J. (1979). Reducing damage to shade and woodland trees from construction activities. *J. Arboricult.* **5**, 97–105.

PART IV

Field and Laboratory Exercises

Simple observations of diseased samples and memorization of facts associated with specific diseases are not sufficient to prepare a student for independent diagnosis. Hands-on experiences are required. In this section we describe several exercises ranging from elementary to advanced, which demonstrate certain disease concepts and which allow students to experience the techniques utilized in disease diagnosis.

A broad selection of exercises is provided to allow choices dependent on facilities, equipment, supplies, and student level. Some can and should be performed by individuals; others can be done in groups; still others can be demonstrated by the instructor.

We have designed the format of each exercise to include an objective or objectives, the anticipated duration, the materials required, and detailed procedures for completing the exercise. General supplies and equipment should include the following: compound and dissecting microscopes, bunsen burners or alcohol lamps, inoculation tools, razor blades, scalpels, sterilizing solution (e.g., 1 part sodium hypochlorite bleach, 1 part 95% ethyl alcohol, 3 parts water), lens paper, forceps, paper towels, glass slides, cover glasses, and lactophenol with and without stain. Specific materials are given with each exercise.

EXERCISE 1

Symptoms and Signs

Objectives: To collect and categorize various symptoms and signs of disease
Duration: One laboratory period
Materials: Pocket knife
Hand lens
Procedures: Spend about half of the laboratory period in the field collecting disease samples. Try to collect diseases that represent a variety of symptoms and signs. Bring the samples back to the laboratory and identify and label the specific symptoms and signs as described in Chapter 7. Check with the laboratory instructor for accuracy of identifications. Examine samples collected by others.
Film: "Examining tissue sections for reproductive structures (signs)." 3 min, 23 sec. Available from APS Headquarters, 3340 Pilot Knob Road, St. Paul, Minnesota

EXERCISE 11

Collection and Identification of Powdery Mildews

Objective: To collect and identify powdery mildews from three different hosts
Duration: 1–2 weeks
Materials: Herbarium paper
Procedures: Obtain leaves showing evidence of powdery mildew from at least three hosts. For each sample remove several cleistothecia and mount them in water. Do not press the cover glass at this point. While observing one or more cleistothecia under low power of a compound microscope, press the cover glass gently with an inoculation needle. Observe if more than one ascus or more than eight ascospores are released by a crushed cleistothecium. Identify the organism to genus by the type of appendages and number of asci present (see Chapter 9, Powdery Mildew). If possible, identify the organisms to species by referring to a host index. Dry the leaf samples and mount individually on herbarium paper. With each leaf, attach a semipermanent slide illustrating the identifying characteristics of the pathogen.

EXERCISE III

Disease Sample Collection

Objective: To collect and identify samples of diseased material and/or disease-causing organisms
Duration: 1–3 months (depending on number of samples to be collected)
Materials: Petri plates with nutrient media
Herbarium paper for mounting leaf samples
Boxes with cubicles for separating large samples or, plywood or peg-board for mounting them
Procedures: 1. Collection. Collect samples of diseased host material or disease-causing organisms from trees and shrubs in fields and forests. Hosts from which the samples are obtained should be living. Each sample should include diagnostic characters which enable one to identify the disease. Where identification requires making microscope slides, slides showing diagnostic features of the disease or disease-causing organism should accompany the sample.

2. Identification. Identify the primary causal agent of each disease sample by looking at the problem from three different levels. This is called the "zoom approach" to disease diagnosis.

a. Macroscopic distant view—This is the bird's eye view from 20 m. Observe the whole host, noting gross abnormal characteristics such as dead twigs, wilted leaves, large cankers or wounds, presence of other organisms, etc.

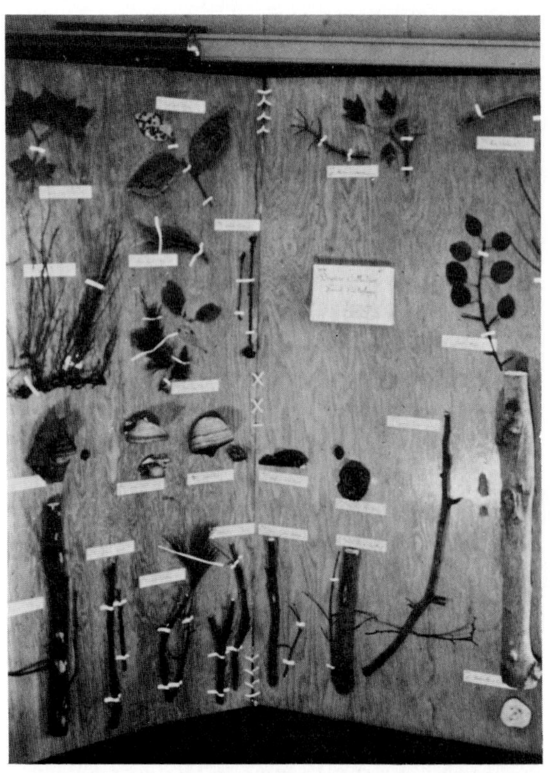

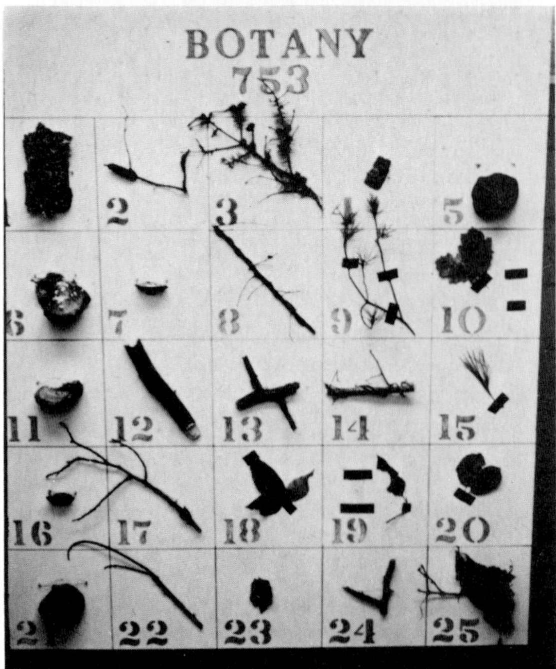

Fig. III.1 Disease sample collections mounted on plywood.

Along with these symptoms and signs, note associated environmental factors such as terrain, buildings, telephone and electric lines, paths, roadways, and other vegetation.

b. Macroscopic close-up view—This is the animal's eye view from 20 cm. Observe abnormal parts of the host with the naked eye and with magnifying lenses of 20× or less. This view should yield information as to the extent of the problem and the spatial relationship of diseased parts to healthy parts. Some small causative microorganisms may be viewed in their fruiting stages at this level.

c. Microscopic view—This is the microorganism's eye view from 20 μm. For isolation and identification of certain causative agents or for detection of physiological and anatomical changes in host tissues, it is often necessary to view the problem with a light microscope. Follow directions for isolating pathogenic organisms given in Chapter 3 and Exercise IV.

3. Mounting. Mount diseased leaves on herbarium paper. Larger samples should be mounted on plywood (Fig. III.1) or pegboard, or stored in boxes with separated cubicles.

4. Labeling. Each sample should have a label with the following information: name of disease, primary causal agent, host, where found, date collected, and name of collector. More elaborate labels could include a brief description of the symptoms and signs and references used to identify the diseases or disease-causing organisms.

Film: "Examination of infected plant tissue and disease diagnosis." 3 min, 23 sec. Available from APS Headquarters, 3340 Pilot Knob Road, St. Paul, Minnesota.

EXERCISE IV

Isolations of Vascular Wilt Pathogens

Objectives: To isolate and identify pathogens from infected woody plant tissues
Duration: 2 weeks
Materials: Living twigs of American elm, approximately ½ inch (1.5 cm) diameter and 6 inches (15 cm) long, which contain abundant brown discoloration in the outer xylem
Two elm twigs from a branch without discoloration
Petri dishes with approximately 25 ml nutrient sterile medium such as potato-dextrose agar or malt extract agar
Procedures: Wipe the bench area with surface sterilant and assemble petri dishes, alcohol burner, and surface sterilant (Fig. IV.1A). Select infected elm twig and dip one end into surface sterilant (Fig. IV.1B). Remove and allow to dry for a few seconds. Cut around the circumference of the twig in the middle and peel off bark with a knife from one end (Fig. IV.1C). Dip knife and forceps into surface sterilant each time before touching the twig. Cut a notch in the twig (Fig. IV.1D) and make a series of six to eight back cuts into the twig moving away from the notch (Fig. IV.1E). Cuts should be deep enough to include some discolored tissue. Break off small chips of wood with the forceps (Fig. IV.1F) and place about seven of them vertically into the agar (Fig. IV.1G). Keep the lid of the petri dish over the agar and only lift the cover high enough to permit placement of the chip into the agar. Store plates at room temperature

Fig. IV.1 Procedures for isolation of *Ceratocystis ulmi*. Assemble petri dishes, alcohol burner, and surface sterilant (A). Select an elm twig and dip one end into surface sterilant (B), then allowing one minute to dry. Sterilize knife in flame and cut around circumference of twig in the middle. Peel bark with thumb and knife from end of twig (C). Cut notch in twig near end (D). Make 6–8 back cuts moving away from the notch (E). Remove chips with sterile forceps (F). Place chips into agar until they touch the bottom of the dish (G). Petri dish one week after incubation of chips at room temperature in the dark (H). Note similar colonies of fungi from most chips, a mixed colony from the uppermost chip and two contaminant colonies between chips near top.

in the dark. In a similar manner prepare one plate from healthy twigs and leave a few plates unopened to serve as controls.

Observe plates after one week and after two weeks (Fig. IV.1H). Make slides of the cultures. Observe cultures directly with the binocular microscope, and make slides of fungal material for examination with the compound microscope. Note any structures that appear around the chips during each observation. Compare these with those illustrated for Dutch elm disease (see Fig. 10.7).

These procedures can be used with any other vascular wilt fungus such as those listed in Chapter 10, Wilt Diseases, or can be modified to enable isolation of fungal or bacterial pathogens from any woody tissues and leaves. Refer to Chapter 3 for specific techniques in isolating leaf pathogens.

Films: 1. "Isolating Pathogenic Fungi." 8 min, 13 sec.
2. "Isolating Pathogenic Bacteria." 11 min, 13 sec.
Available from APS Headquarters, 3340 Pilot Knob Road, St. Paul, Minnesota.

EXERCISE V

Identification of A and B Mating Types of *Ceratocystis ulmi*

Objective: To determine the mating types of unknown isolates of *C. ulmi*
Duration: 3 weeks to 3 months
Materials: Cultures of A and B mating types of *C. ulmi*
Cultures of unknown mating types of *C. ulmi* (number can vary depending on availability and desired sophistication of the exercise
Elm branches, 1½ to 2½ inches in diameter
Glass petri dishes

Procedures (Modified from Tuite, 1969): Remove the bark from the elm branches and cut the branches into ⅜ inch discs. Put one disc on 2–3 filter papers in a petri plate. Add 20 ml of tap water and autoclave at 15 psi for one hour. After the petri plates have cooled, place a small amount of mycelium and agar of mating type A on the wood disc. Place a similar amount of mycelium and agar from an unknown mating type about 1 inch from the known sample. Repeat the procedure with mating type B and the unknown. Test as many unknowns (two replications per isolate) as desired in the same manner. Known mating types A and B should be tested together as a control. Incubate at 15–20° C. Compatible types will produce perithecia beneath the mycelial growth sometime between 3 weeks and 3 months. Periodic checks should be made for the black, long-necked perithecia. Record presence or absence of perithecia and determine the mating types of the unknowns, i.e., an unknown will be mating type A if it produces perithecia with known mating type B and vice versa. If perithecia do not develop with either A or B mating types, the unknown mating type cannot be determined.

Literature Cited

Tuite, J. (1969). "Plant Pathological Methods." Burgess, Minneapolis, Minnesota.

EXERCISE VI

Koch's Postulates

Objectives: To review principles of pathogen isolation, inoculation of a healthy host, development of disease, and reisolation of a pathogen from a diseased host

Duration: 2–4 weeks after host inoculation, but seedlings must be maintained 3–6 months prior to inoculation.

Materials: Potted seedlings of American elms

Agar cultures of *Ceratocystis ulmi,* isolated from an elm tree with Dutch elm disease (can be the organism isolated in Exercise IV

Parafilm

Reisolation of *C. ulmi* requires the same materials as needed in Exercise IV

Procedures: Make a shallow downward cut into the lower stem of an elm seedling making sure xylem is cut. Place a small amount of an active *C. ulmi* culture from an agar plate into the bark flap made by the wound. Wrap the entire stem around the wound with parafilm. Record the occurrence of any symptom expression over the next 2–4 weeks. When obvious wilt symptoms appear, the stem should be sectioned into 6 inch (15 cm) pieces and isolations made for wilt pathogens (see Exercise IV). Identify the pathogen isolated. To serve as controls, two trees should be inoculated with agar from sterile plates and two trees should be unwounded.

Additional Note: Although these procedures can be adapted for use in other host–pathogen combinations, *C. ulmi–U. americana* was chosen because of ease in successful inoculation, rapid disease progression, and lack of juvenile resistance problems in the host.

EXERCISE VII

Extraction of Nematodes from Soil and Roots*

INTRODUCTION AND OBJECTIVES

Before nematodes can be examined for identification, they must be separated from soil particles. An ideal method would consistently remove most of the nematodes from a large soil sample including all of the species present. It should be quick, easy, and use simple equipment. The nematodes recovered should be in good shape anatomically and physiologically.

Two methods currently in use will be described. They are both dependent on various characteristics of nematodes: their ability to move, their large size in relation to soil particles, and their low specific gravity.

BAERMANN FUNNEL (Fig. VII.1)

Principle: Nematodes will move downwards through a thin layer of saturated soil held on a porous paper or cloth. Nematodes cannot swim and will settle to the bottom of a container.

*From R. A. Rohde, University of Massachusetts, Amherst.

Fig. VII.1 Baermann funnel. The soil or root sample is placed on wet strength tissue which is supported by an aluminum platform (A). The water level in the funnel is kept in contact with the soil or root sample just enough to keep it moist but not flooded. A petri dish cover is often placed over the top of the funnel to retard sample drying (B).

Materials: A glass funnel terminating in a length of rubber tubing sealed with a clamp
An aluminum wire platform which will rest ⅓ of the way down the funnel cone

Procedures: 1. Place a layer of wet strength tissue over the screen. Add 50 ml of soil in a thin layer.

2. Fill the funnel carefully with water until the sample is saturated by capillarity.

3. After 24–48 hours, open the clamp and draw off 10 ml of water and examine for nematodes. Migration will continue for several days, but nematodes deteriorate unless drawn off each day.

SUGAR FLOTATION

Principle: Nematodes will float on a sugar solution in which soil particles sink.

Materials: Clinical centrifuge, swinging bucket, for 50 ml tubes (Fig. VII.2)
Sugar syrup—dissolve 1 pound cane sugar in one liter of hot water, let cool
Five gallon pail
325 mesh sieve
Household strainer

Procedures: 1. Place 250 ml of soil in the strainer, and wash with water into the pail. This sieve separates the coarsest debris. The pail should be ½ to ⅔ full.

Fig. VII.2 Table top clinical centrifuge suitable for sugar flotation preparation of nematodes.

2. Stir water in the pail thoroughly. Allow to settle for 30 seconds. Decant through a 325 mesh sieve.

3. Wash sieve residue into a 250 ml beaker. Aim for a volume of 200 ml.

4. Pour into 4 polypropylene tubes, stirring the beaker constantly to keep the soil suspended. This equalizes weight.

5. Centrifuge at 3450 rpm (full speed) for 4 minutes.

6. Pour off supernatant. Retain the pellet.

7. Fill each tube ⅔ full with sugar solution. Stir thoroughly with a spatula to break up the pellet.

8. Centrifuge at 3450 rpm for 30 seconds.

9. Pour supernatant onto a clean 325 mesh sieve. Wash thoroughly to remove sugar. Wash residue containing nematodes from the sieve into a beaker or syracuse dish.

10. Observe nematodes in beaker.

Film: "Separating Nematodes from Soil." 10 min, 43 sec. Available from APS Headquarters, 3340 Pilot Knob Road, St. Paul, Minnesota.

EXERCISE VIII

Observations of Mycorrhizae

Objectives: To observe mycorrhizae on roots of trees, and to differentiate between roots with ectomycorrhizae and roots with VA mycorrhizae

Duration: One laboratory period

Materials: Seedlings (*Pinus* spp. and *Acer* spp.)
Petri dishes
Dilute hydrochloric acid
10% potassium hydroxide (KOH)
0.05% Trypan blue in lactophenol

Procedures: Remove the soil ball of the seedlings from their containers. Carefully and thoroughly wash the soil from around the roots in a dirt sink or bucket. Cut sections of the small secondary roots of each tree and place them separately in petri dishes partially filled with water. Observe the small feeder roots with a dissecting microscope at a range of magnifications and compare with drawings and photographs of mycorrhizae (see Figs. 13.13, 13.14, 13.15).

Staining of VA mycorrhizae (Phillips and Hayman, 1970). Heat several sections of secondary feeder roots (approximately 2 mm in diameter and 5 mm long) at 90°C in 10% KOH for 1–2 hours to clear the sections for viewing. Rinse the sections in water and acidify in dilute HCl. Place in 0.05% Trypan blue stain in lactophenol and simmer for 5 minutes. Move sections to clear lactophenol to remove stain and mount on glass slides with coverslips to observe under a compound microscope. This procedure may need to be modified for each species depending on the amount of pigmentation of the roots and the thickness of the roots and the thickness of the sections used.

Additional Note: Any other tree species that exhibit easily seen ectomycorrhizae or VA mycorrhizae could be substituted for the ones used in this exercise.

Literature Cited

Phillips, J. M., and Hayman, D. S. (1970). Improved procedures for clearing roots and staining parasitic and vesicular mycorrhizal fungi for rapid assessment of infection. *Trans. Br. Mycol. Soc.* **55**, 158–161.

EXERCISE IX

Development of Crown Gall

Objectives: To observe the development of the crown gall disease in tree seedlings

Duration: 2 months

Materials: Willow seedlings (*Salix* spp.)
Culture of *Agrobacterium tumefaciens*
Parafilm
50 ml beaker
Eye dropper

Procedures: With a scalpel, wound seedlings on the upper stem with a downward cut to the xylem. Remove the petri dish cover and wash the *A. tumefaciens* culture into a 50 ml beaker with approximately 25 ml of water. Add about 1 ml of the bacterial suspension to the wound with an eye dropper. Wrap parafilm around wounds and remove in 1 week. To serve as controls, leave one wounded tree uninoculated, smear a loop of bacteria on the base of an unwounded tree, and leave one tree uninoculated and unwounded. Observe and note changes in the trees during the next 2 months. At the end of the exercise, section through the stem and gall tissue. Observe the anatomical changes that occurred and compare with healthy stem tissue.

EXERCISE X

Evaluation of Shade Trees

Objective: To determine the monetary value of shade trees over 12 inches in diameter

Duration: One laboratory period

Materials: Pocket calculator

Procedures: The International Society of Arboriculture (1975) has published a guide which explains a method for determining the value of trees on the basis of the following criteria:

a. Number of square inches in cross section of the tree trunk at 1.5 m above-ground

b. Standard value at $18 per square inch (this value is current as of 1979, but is periodically revised).

c. Species group value in percent:
   Group I, 100%; Group II, 80%; Group III, 60%; Group IV, 40%;
   Group V, 20%.

d. Physical condition in percent; five categories:
   I. perfect, 100%; II. 80%; III. 60%; IV. 40%;
   V. least perfect, 20%.

e. Aesthetics (location on and enhancement of property) in percent.

Numbers representing these criteria are put into a formula, and the final value of a tree is determined. The formula is:

$$a \times b \times c \times d \times e = \text{final value}$$

Select trees from one or more of the species groups below which can be found within walking distance, and determine the final value(s) using the above formula. Compare results with those determined by others.

Group I	Group II	Group III
Red maple	Norway maple	Gray birch
Sugar maple	Shagbark hickory	Red pine
American beech	White pine	American linden
White oak	Green ash	Colarado blue spruce
American elm	Sycamore	

Group IV	Group V
Norway spruce	Boxelder
Silver maple	Most poplars
Mulberry	Most willows
Common horsechestnut	Most alders

Literature Cited

International Society of Arboriculture (1975). "A Guide to the Professional Evaluation of Landscape Trees, Specimen Shrubs, and Evergreens," Revision 3. Urbana, Illinois.

EXERCISE XI

Effects of Salt on Tree Seedlings

Objective: To compare the effects of soil-applied versus foliar-applied NaCl at various concentrations on selected species of seedlings

Duration: 3–5 weeks

Materials: Seven three-year-old seedlings from each of at least two selected coniferous species, potted in identical soil, and growing under greenhouse conditions. Younger, foliated, deciduous species can also be used

NaCl concentrations in distilled water of 500, 1000, and 2000 ppm (w/v)

Mist sprayer with capacity of at least 100 ml

Procedures (Modification from Costantini and Rich, 1973): Select one tree from each species tested as a control. Treat each of the remaining six trees of each species as follows:

Tree #	Treatment (applied once daily for 5 days)
1	Foliar spray, 100 ml, 500 ppm NaCl
2	Foliar spray, 100 ml, 1000 ppm NaCl
3	Foliar spray, 100 ml, 2000 ppm NaCl
4	Soil application, 100 ml 500 ppm NaCl
5	Soil application, 100 ml 1000 ppm NaCl
6	Soil application, 100 ml 2000 ppm NaCl

The foliar spray should be evenly distributed throughout the foliage, and the soil application should be evenly distributed over the soil surface. Controls should be treated with an equal volume of water. Observe the seedlings at selected intervals for foliar symptoms. Compare species differences and method of NaCl application.

Literature Cited:

Costantini, A., and Rich, A. E. (1973). Comparison of salt injury to four species of coniferous tree seedlings when salt was applied to the potting medium and to the needles with or without an antitranspirant. *Phytopathology* **63,** 200 (Abstr.)

EXERCISE XII

Electrical Detection of Disease in Living Trees

Objective: To detect and monitor the presence and progression of disease in living trees with electrical devices

Duration: One laboratory period to several weeks

Materials: Shigometer, if available

Multimeter that will measure electrical resistance in ohms (may be used in place of a Shigometer, but without the accuracy or stability of that instrument)

Needles, nails, pins, or other sharpened metal pieces that can be used as electrodes for insertion into woody tissues

Alligator clips

Electrical hook-up wire

Two-foot cross section of a freshly cut tree exhibiting internal discoloration and decay

Twenty elm seedlings approximately 1 m in height

Viable culture of *Ceratocystis ulmi*

Procedures: Before performing the procedures described in this exercise and in Exercise XIII, the experimenter should read the selected references at the end of this exercise to become acquainted with the basic concepts involved. It should be understood that the procedures described here are for comparative measurements, since a wide variation in electrical characteristics exists between individual trees.

1. Discoloration and decay—Insert two similar-metal electrodes into decayed tissue of a freshly cut tree sample. The electrodes should be 2 cm apart and inserted to a depth of 2 cm. Attach the electrodes to the input of a Shigometer or multimeter. Record the electrical resistance (ER) in thousands of ohms (kohm). Remove the electrodes and insert them in discolored wood in a manner as just described. Discolored wood is found between decayed wood and healthy wood. Record the electrical resistance. Repeat the exercise for clear, healthy wood. Compare the ER measurements for the three tissues. Decayed wood should show the least ER, healthy wood the most. This procedure works on the principle that decayed and discolored tissues offer less resistance to the flow of electrical current generated in the Shigometer or the multimeter.

2. Presymptomatic detection of stress in living trees—Insert two similar-metal electrodes through each stem of several elm seedlings (Fig. XII.1). Place the electrodes 10–20 cm apart in a vertical direction and about 0.3 m above the

Fig. XII.1 Elm seedling with inserted pin electrodes connected to an electrical resistance meter.

Fig. XII.2 Liquid-tight parafilm boat around base of elm seedling. Note pin electrodes above boat.

soil line. Allow two days for the electrodes to equilibrate. Record the electrical resistance between the electrodes in each seedling with a Shigometer or a multimeter. Divide the seedlings into a control group and a test group such that the mean ER for each group is the same. Submit the test trees to one of the following stress conditions or to other stress conditions of choice:

a. Inoculate the seedlings with viable spores of *Ceratocystis ulmi.* This can be done by wounding the stem a few centimeters above the soil line and by bathing the wound with a liquid suspension of the pathogen.
b. Attach a liquid-tight parafilm boat around the base of the stem (Fig. XII.2). Fill the boat with one of various toxic substances, such as 5–10% HCl or 1–3% catechol. With a scalpel, wound the stem below the surface of the liquid (Fig. XII.3).
c. Remove several or all of the leaves.
d. Induce drought by limiting water.
e. Keep the soil saturated by immersing the pot in water.

Fig. XII.3 Scalpel wound in elm stem below liquid surface.

Take ER readings from each seedling daily for the duration of the exercise. Readings should be done at the same time each day. Record visual symptoms when and if they occur. Compare the mean ER of the control group with the mean ER of the test group over time. Compare changes in ER with symptom expression.

Selected References

Blanchard, R. O., and Carter, J. K. (1980). Electrical resistance measurements to detect Dutch elm disease prior to symptom expression. *Can. J. For. Res.* **10,** 111–114.

Blanchard, R. O., and Shortle, W. C. (1977). Changes in electrical resistance associated with disease and death of elm seedlings. *Proc. Am. Phytopathol. Soc.* **4,** 183. (Abstr.)

Shigo, A. L., and Shigo, A. (1974). Detection of discoloration and decay in living trees and utility poles. *USDA For. Serv., Res. Pap. NE* **NE-294.**

Skutt, H. R., Shigo, A. L., and Lessard, R. A. (1972). Detection of discolored and decayed wood in living trees using a pulsed electric current. *Can. J. For. Res.* **2,** 54–56.

Tattar, T. A. (1974). Measurement of electric currents in clear, discolored, and decayed wood from living trees. *Phytopathology* **64,** 1375–1376.

Tattar, T. A. (1976). Use of electrical resistance to detect Verticillium wilt in Norway and sugar maple. *Can. J. For. Res.* **6,** 499–503.

Tattar, T. A., and Saufley, G. C. (1973). Comparison of electrical resistance and impedance measurements in wood in progressive stages of discoloration and decay. *Can. J. For. Res.* **3,** 593–595.

Tattar, T. A., Shigo, A. L., and Chase, T. (1972). Relationship between the degree of resistance to a pulsed electric current and wood in progressive stages of discoloration and decay in living trees. *Can. J. For. Res.* **2,** 236–243.

Wargo, P. M., and Skutt, H. R. (1975). Resistance to a pulsed electric current: An indicator of stress in forest trees. *Can. J. For. Res.* **5,** 557–561.

EXERCISE XIII

Construction of a Discoloration and Decay Detector

Objective: To build an inexpensive detector for measuring comparative electrical differences between healthy, discolored, and decayed wood
Duration: 1–2 weeks
Materials: Microammeter (0–50 microamperes, direct current (DC). 1½ volt D battery
Battery holder
Electrodes as in Exercise XII, except that one electrode should be of a different kind of metal than the others
Alligator clips
Hook-up wire
Several ½ watt resistors ranging from 5000–25,000 ohms
Tree section as in Exercise XII

Procedures: 1. Battery generated current (Fig. XIII.1). Place the D battery in the battery holder. With alligator clips and hook-up wire, attach the negative side of the battery to the negative side of the microammeter. Attach the positive side of the microammeter to one of the ½ watt resistors. The value used will depend on how much decay is present. If, when making measurements, the meter needle goes off scale, use a resistor of higher value. If the meter needle moves very little, use a resistor of lower value. Connect the positive side of the battery to one electrode and the free end of the resistor to the other electrode.

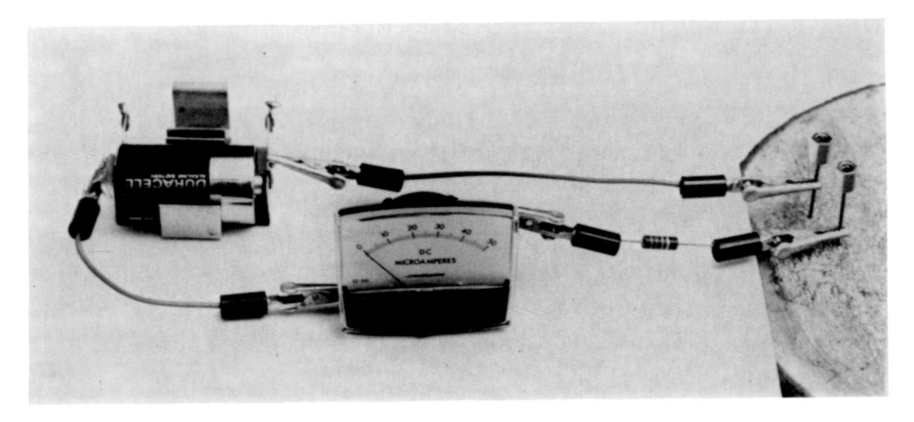

Fig. XIII.1 Arrangement of components for measuring relative resistance to battery generated electrical current through healthy, discolored, and decayed wood.

Both electrodes should be of the same type of metal. Insert the electrodes into the various tissues as described in Exercise XII. Note comparative differences in readings on the microammeter. Decayed tissue has less resistance to current flow than discolored or healthy tissue. Therefore, more current will pass through decayed tissue and will be detected as a higher reading in microamperes on the microammeter. The difference between this detector and a Shigometer or multimeter is that the circuitry of the latter two instruments

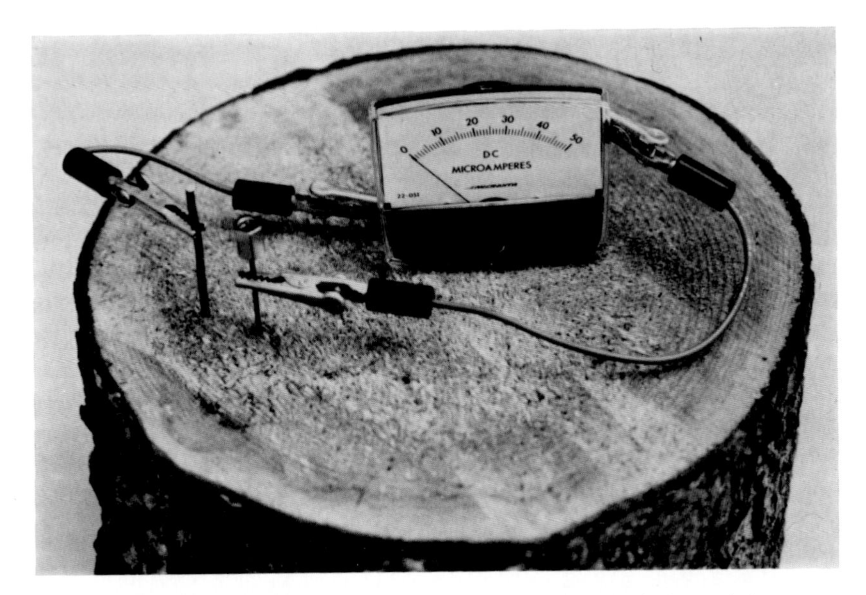

Fig. XIII.2 Arrangement of components for measuring relative production of tissue generated electrical current in healthy, discolored, and decayed wood.

transcribes current flow into a measurement of electrical resistance, which is inversely proportional to current.

2. Tissue generated current (Fig. XIII.2). Insert two dissimilar-metal (e.g., copper and zinc) electrodes into the tree sample used in procedure 1 above. Depth and distance apart should be the same as described in procedure 1. Connect the electrodes to the microammeter. Record the electrical current in microamperes. Note that more current will be generated if the electrodes are inserted deeper into the tissues. Decayed wood should show the most electrical current, healthy wood the least. This procedure works on the principle that two dissimlar metals inserted into an electrolyte (in this case, the woody tissue) will act as a battery, and electrical current will be generated.

Glossary

Abiotic Nonliving
Acervulus Mat of hyphae, generally associated with a host, forming erumpent lesions with short, densely packed conidiophores
Aecium cuplike fruiting structure of the rust fungi
Anthocyanescence purplish or reddish coloration of leaves or other organs due to overdevelopment of anthocyanin pigment
Apothecium open, cuplike ascocarp
Ascocarp sexual fruiting body of the Ascomycetes containing one or more asci
Ascospore spore produced in an ascus
Ascus saclike structure containing ascospores formed as a result of karyogamy and meiosis
Autoecious completing the entire life cycle on one host; generally applied to rust fungi
Bacterium typically unicellular microorganism without chlorophyll, dividing by fission; nuclear material is not surrounded by a nuclear envelope
Basidiocarp sexual fruiting body of the Basidiomycetes containing basidia
Basidiospore spore produced on a basidium
Basidium generally club-shaped structure on which basidiospores are produced as a result of karyogamy and meiosis
Biotic living
Blight rapid killing of foliage, blossoms, and twigs
Blotch large, irregular lesions on leaves, shoots, and stems
Callus overgrowth of tissues at the margins of wounds and diseased tissues
Canker necrotic, often sunken lesion in the cortical tissues of stems and roots
Chlamydospore thick-walled asexual resting spore, formed from a vegetative cell
Chlorosis failure of chlorophyll development in normally green tissues

Cirrhus twisting tendril of spores held together by mucus as it passes through an ostiole
Clamp connection bridgelike hyphal connection characteristic of many Basidiomycetes
Cleistothecium closed ascocarp
CODIT an achronym for compartmentalization of decay in trees
Coenocytic multinucleate
Conidiophore specialized hyphal branch on which conidia are produced
Conidium asexual spore formed on a conidiophore
Conk Woody shelflike basidiocarp characteristic of many wood-rotting fungi
Curl rolling or folding of leaves due to localized overgrowths of tissue
Decay disintegration of dead tissues
Diagnosis identification of disease from its symptoms and signs
Dieback progressive death of twigs and branches from their tips toward the trunk
Dikaryotic having two haploid nuclei per cell
Diploid having nuclei with $2n$ number of chromosomes
Disease any deviation from the normal state of an organism that impairs its vital functions, developing in response to genetic defects or to an unfavorable environmental factor
Dolipore septum specialized septum found in many Basidiomycetes with a central pore surrounded by a barrel-like swelling of the septal wall and covered by a perforated membrane
Dwarfing subnormal size in an entire plant or some of its parts
Erumpent bursting through the surface of a substratum
Etiolation yellowing due to lack of light
Fasciation flattened condition of a plant part that is normally round
Fasciculation broomlike growth of densely clustered branches, often referred to as witches' broom
Felt densely woven mat of mycelium
Flagellum appendage giving motility to a cell
Flux slimelike material flowing from a tree and containing bacteria, fungi, and other microorganisms
Fruit body a reproductive structure on or in which spores of a fungus are produced
Fumigant gaseous compound lethal to insects and fungi
Fungicide substance that kills fungi
Fungus achlorophyllous, spore-bearing eukaryote with a walled thallus and absorptive nutrition
Fusiform spindlelike
Gall swelling or outgrowth on a plant, caused by a pathogen
Germination to begin to grow as a seed or spore
Germ tube the first hypha emerging from a germinating spore
Gill lamellate structure on the underside of a mushroom cap
Haploid having nuclei with $1n$ number of chromosomes
Haustorium absorbing organ of a fungus which penetrates a host cell without penetrating the plasma membrane
Heteroecious requiring two unrelated hosts to complete the life cycle; generally applied to rust fungi
Host living organism upon which another organism grows and obtains all or part of its food
Hyaline lacking color
Hydrosis water-soaked, translucent condition of tissues due to cell sap passing into intercellular spaces
Hyperplasia overgrowth due to an increase in number of cells
Hypertrophy overgrowth due to an increase in size of cells
Hypha Tubular, branching filament of a fungus thallus
Hypovirulent condition of subdued virulence of a pathogenic strain
Hysterothecium elongated, boat-shaped apothecium with a longitudinal slit
Incubation period between inoculation and appearance of visible symptoms
Infection establishment of a food relationship between a parasite and a host

Infestation intermixing of one organism with another without establishing a food relationship
Inoculate bringing a pathogen to a portal of entry of the host
Inoculum any part of a pathogen capable of growing and causing infection
Intercellular between cells
Intracellular within cells
Karyogamy fusion of nuclei or nuclear material
Lesion usually a local, well-defined, diseased area
Macerate to soften or separate into constituent parts
Macroconidium large conidium of fungi which have conidia of two distinct sizes
Meiosis reduction division of $2n$ parent nuclei to yield $1n$ daughter nuclei
Microconidium small conidium of fungi which have conidia of two distinct sizes
Mildew cobwebby or powdery superficial growth, usually on leaves
Mitosis division of parent nuclei to yield daughter nuclei of the same chromosome number
Mold woolly or furry surface growth of mycelium
Mushroom (=Toadstool) umbrella-shaped fruiting structure of many Basidiomycetes
Mycelium a mass of fungal threads or hyphae
Mycorrhiza structure resulting from a symbiotic relationship between fungal mycelia and the roots of trees and other higher plants
Necrosis death
Nucleus part of a cell primarily made up of chromosomes
Obligate generally refers to parasites which must grow on or in living cells
Oospore sexual resting spore resulting from the union of unlike gametes
Ooze viscid mass made up of plant juices and often pathogen cells
Ostiole a pore through which spores are released, generally associated with perithecia, pycnia, and pycnidia
Parasite an organism that lives on or in another organism
Pathogen any factor or agent capable of causing disease
Perithecium characteristically flask-shaped, ascus-containing fruiting body with an ostiole and a wall of its own
Phylogeny study of the evolutionary history and interrelationships of various species
Plasmodium naked mass of protoplasm
Plasmogamy fusion of cytoplasm of two cells
Predisposition environmental modification of resistance barriers making plants more susceptible to pathogens
Propagule an organism or part of an organism capable of generating a new organism
Prophylaxis prevention
Pseudothecium fruiting body bearing asci in locules within a stroma
Pycnidium asexual, hollow fruiting body containing conidia
Pycniospore spore borne in pycnia of the rusts which acts as a male gamete, fusing with receptive hypha
Pycnium fruiting structure of the rusts in which are produced receptive hyphae and pycniospores
Receptive hypha hypha protruding from a pycnium and acting as a female gametangium in the rusts
Rhizomorph cordlike strand of fungal hyphae
Rosetting crowded condition of foliage due to lack of internode elongation
Saprophyte an organism that lives on dead organic matter
Sarcody abnormal swelling of tissues above girdled branches or stems
Scab roughened, crustlike lesion
Scald blanching of the epidermis and adjacent tissues
Sclerotium hard, compact, resting body composed of fungal hyphae
Scorch browning of leaf margins from death of tissues
Septum cross-wall dividing two cells

Shot-hole circular holes in leaves resulting from the dropping out of the central necrotic areas of spots
Sign structure of a pathogen, usually signifying a disease
Sorus mass or cluster of spores borne on short stalks
Spermatization plasmogamy between a pycniospore and a receptive hypha in the rusts
Sporangiospore asexual spore borne within a sporangium
Sporangium enlarged tip of specialized hyphal branch in which sporangiospores are produced
Spore general name for a single to several celled progagative unit in the fungi and other lower plants
Sporodochium cushion-shaped stroma covered with conidiophores
Spot lesions, usually defined, circular or oval in shape, with a central necrotic area surrounded by variously colored zones
Stipe stem of a mushroom
Stroma compact mass of fungal hyphae on or within which fruiting structures are formed
Suppression prevention of the development of certain organs
Symbiosis a living together of two dissimilar species, often for mutual benefit
Symptom any condition in a host resulting from disease that indicates its occurrence
Teliospore in the rusts, the spore in which karyogamy and meiosis occur
Telium sorus containing teliospores
Thallus vegetative phase of a fungus
Therapy treatment of disease by application of chemicals or heat to eliminate the pathogen or application of elements to cure deficiency diseases
Tumefaction tumorlike or gall-like overgrowth of tissue
Urediospore repeating spore of the rusts
Uredium sorus containing urediospores
Vector an animal, often an insect that disseminates pathogens
Virescence development of chlorophyll in tissues where it is normally absent
Virulence degree of pathogenicity
Virus submicroscopic disease-causing agent which multiplies only in living cells
Wilt leaves or shoots lose their turgidity and droop
Witches' broom see Fasciculation
Yellowing leaves turn yellow due to a degeneration of chlorophyll
Zoospore motile spore

Index